WOMAN HERSELF

To all the women I love,
and to those who join us
in moving passionately forward

'Woman herself, must judge of woman'

Matilda Joslyn Gage
Woman, Church and the State (1893)

WOMAN HERSELF

a transdisciplinary perspective
on women's identity

Robyn Rowland

OXFORD
UNIVERSITY PRESS

Melbourne
Oxford   Auckland   New York

OXFORD UNIVERSITY PRESS AUSTRALIA

Oxford   New York   Toronto
Delhi   Bombay   Calcutta   Madras   Karachi
Petaling Jaya   Singapore   Hong Kong   Tokyo
Nairobi   Dar es Salaam   Cape Town
Melbourne   Auckland
and associated companies in Berlin   Ibadan

OXFORD is a trade mark of Oxford University Press

National Library of Australia
Cataloguing-in-Publication data:

Rowland, Robyn, 1952—   .
   Woman herself: a transdisciplinary
   perspective on women's identity.

   Bibliography.
   Includes index.
   ISBN 0 19 554475 7.

   1. Women.   2. Identity (Psychology).
   I. Title.

305.4′2

Edited by Sarah Brenan
Designed by Steven Randles
Typeset by Setrite Typesetters Ltd, Hong Kong
Printed in Australia by The Book Printer, Victoria
Published by Oxford University Press,
253 Normanby Road, South Melbourne, Australia

Contents

Acknowledgements

This has been a difficult book to write, believe me. It has been with me for over ten years in concept and six years in the writing. It has become more than the Women's Studies text it was intended to be: it is my statement on the oppression of women and our passionate resistance to it. It is a contribution to feminist revolutionary change, written in the hope that it will bring women closer to our Selves and to other women, the personal and political embracing, empowering us to act politically to create a world where women's oppression can be taught as history, not lived daily in bone and flesh.

So many years have brought so much support and encouragement that my thanks can only touch this. A filigree of love has been twined for me by my women friends. Renate Klein understands me. Her encouragement and faith sustain me always; her love is a bedrock of my life. We share an intellectual companionship, a commitment to women, a love of diversity, and a stubbornness of belief. She critically read this manuscript *before* it was cut, a monumental effort which only a woman with her generosity and kindness would undertake. Dale Spender has given me, as she has so generously given to so many women, an awe for women's capacity, a confidence to work at what we believe in, and the joy of laughter as a revolutionary act. Lynne Spender has held me up when the seas were very rough, even when she herself was only treading water. And we sing well together!— though too rarely. Zoë Fairbairns, Ailbhe Smyth and Lariane Fonseca I thank for their love and, particularly, for their irreverence!

My parents are both impossible, loving and loveable people. Gwen's emotional strength and courage are inspiring. She taught me not to give *in*—and a whole lot more. Norm taught me not to give *up* and how to be stubborn—I mean persistent! I love them both and thank them for a closeness that still nurtures me.

I thank Nigel Wood, with whom I share the pleasures and difficulties of daily loving, and a commitment to our work. A man of integrity, caring, calm strength, and of political values honestly lived, he still teaches me the names of trees.

My editors have been wonderful. Louise Sweetland amazed me because she never gave up; against all odds she kept believing that I'd deliver. Carla Taines made me cut the book *in half*—to its benefit— and Sarah Brenan helped me polish the remainder.

There were others along the way—among them TQI—who parted company with me but were loving supporters and are not forgotten.

To those who helped me with the nitty-gritty of library work (Rose Mildenhall, Angela Coughlan among them) and the keyboard experts (Jane Veale, Bev Bartlett, Karen Pantony, Jan Wapling, Lee McKay and Chris Wood) as I've told you many times, a *million* thanks. Finally, to you Judy Barber, who orchestrates the experts, who feeds me with some of the most useful critical comments I ever receive, and whose judgement I respect, my thanks and enduring affection.

Introduction

In the last two decades, the women's liberation movement has attacked the artificial division between private and public social domains, that relegates women to the domestic, family sphere, and men to the world of paid work and power. In doing so, feminists have stressed that the experience of the individual woman is often shared by other groups of women, making what were previously private and personal issues social and political.

The knowledge of the political dimension in personal experience has been empowering for women in their struggle for self-respect, integrity, autonomy and power. There are many feminisms operating in the women's liberation movement, and each has differing explanations of the sources of women's oppression and different strategies for changing that injustice. But all agree that this oppression and subordination of women as a social group exists and must be ended.

This book deals with a topic which is simultaneously social, political and personal. The processes of construction of identity are social, in that institutions in society (such as the family) and ideological values (such as those associated with motherhood) combine to inform the individual about 'appropriate' sex-determined behaviour. The way that 'appropriate' behaviour is inculcated or enforced is political— an expression of power relations in society—as are the definitions themselves. Identity can also be self-generated, or transformed, in that the individual, enlightened as to the form and purpose of these social and political constructions, can struggle to reject them and remake or remould her self. This understanding can lead to collective struggle to change the definition of self for women and the social and political structures which create it.

Thus personal *is* political; but raw experience is not enough to lead to cogent social analysis. Feminism needs both, Monique Plaza neatly cojoins the two when she writes:

When they discover diverse systems of exploitation and alienation of which they are the victims, women often ask themselves these questions: 'What are we? What would we be without this social mould? What, basically, is a woman?'[1]

As Gerda Lerner has written, women's 'own definitions of selfhood and fulfilment have remained subordinate to patriarchal concepts'.[2] Women can begin to move towards a feminist definition of woman by unravelling the patriarchal controls placed upon her. To be independent of others' definitions of themselves, to be full of self-knowledge and self-respect, should serve as a solid base for action in their own individual interests, and group action in the interests of other women.

This book synthesizes material which is relevant to an understanding of the self for women in the hope that it will validate personal experience and expose the construction of negative selfhood which imprisons women.* I have selected self-identity as the framework because for many women this is their first and primary reality.

I argue that men have created an identity for women, based in biology, which is intended to reinforce difference and to tie women to a 'natural' position in such a way as to make woman the negative or 'other'. Through patriarchy men direct and try to impose this self on woman for the purposes of controlling her and maintaining women as a servicing class for men. Part of the feminist struggle has been resistance to this imposition of negative selfhood and I will also consider examples of these acts of resistance.

Patriarchy as context

The social and political context in which the construction of self takes place for women is a male-dominated society in which women are oppressed by men—that is, through patriarchy. By patriarchy Rosaldo means a socio-political system in which women are universally subordinate to men; men dominate due to their control of public life and their relegation of women to the domestic, and this differential participation of the sexes in public life gives rise to universal male authority over women and a higher valuation of things male.[3]

In this book patriarchy is used to denote the system of structures and institutions created by men in order to sustain and recreate male power and female subordination. Such structures include institutions such as the law, the economic system, religion, education and the family; ideologies which perpetuate the 'naturally' inferior position of women; and socialization processes which ensure that women and men develop behaviours and belief systems appropriate to the powerful or powerless group to which they belong.

---

* As will become evident, the focus is primarily on the experiences of white women in Western countries, particularly Australia, North America and England.

Patriarchy is ahistorical in the sense of its constancy through history, but it has expressed itself differently at various historical periods, and is experienced differently by different groups of women, for example according to class, race, culture, age and sexual preference. Ruth Bleier draws attention to this:

By patriarchy I mean the historic system of male dominance, a system committed to the maintenance and reinforcement of male hegemony in all aspects of life—personal and private privilege and power as well as public privilege and power. Its institutions direct and protect the distribution of power and privilege to those who are male, apportioned, however, according to social and economic class and race. Patriarchy takes different forms and develops specific supporting institutions and ideologies during different historical periods and political economies.[4]

Through the control of resources, the continuance of male-dominated socialization processes and sometimes the use of force or coercion, men as a group oppress women.

Bell Hooks describes oppression as the 'absence of choices' and maintains that this differentiates between the oppressed and the oppressor. But women who do experience some choice and control in their lives can then mistake it for real liberation. She warns:

Many women in this society do have choices (as inadequate as they are) therefore exploitation and discrimination are words that more accurately describe the lot of women collectively in the United States . . . they may know they are discriminated against on the basis of sex, but they do not equate this with oppression . . . the absence of extreme restrictions lead many women to ignore the areas in which they are exploited or discriminated against; it may even lead them to imagine that no women are oppressed.[5]

Jo Freeman delineates two aspects of oppression: the socio-structural aspects such as legal, economic, social, and political institutions, and the socio-psychological aspects, such as the group self-hate of women. Freeman describes the oppression of women as sexism, which has two core components. First, there is the belief that men are more important, significant, valuable and worthwhile than women, and that the work of men in the world, and hence their time, is more important. Second, there is a belief that women are 'complementary' to men: that they exist for their assistance and pleasure.[6] Women are thus defined in relation to men. Their identity is dependent on the man they are connected with, and women who do not accept this are a basic threat to the social values of patriarchy. Those who choose not to be defined in relation to men are called 'man-haters'.

Difference and sisterhood

Patriarchal power rests in hierarchical structures of oppression which affect different groups of women differently. Primary examples of these structures are capitalism, with its class differentiations, and racism. As Charlotte Bunch has written,

Women's oppression is rooted both in the structures of our society, which are patriarchal, and in the sons of patriarchy: capitalism and white supremacy. Patriarchy includes not only male rule but also heterosexual imperialism and sexism; patriarchy led to the development of white supremacy and capitalism. For me, the term patriarchy refers to all these forms of oppression and domination, all of which must be ended before all women will be free.[7]

Capitalism is a system based on the accumulation of private profit through the ownership by a few of the means of production. Inherent in this system is the exploitation of the wage-labourer. Women have been essential in maintaining capitalism through their waged labour, and their unwaged work which includes reproduction and servicing of the male workforce.

Class will define their opportunities for thought and development, as well as placing material limitations on their lives, and it changes the meaning of the universality of being 'woman'. Women do have a stake in overturning the class system. Until all forms of oppression and exploitation are eliminated, women as a social group cannot be free. In Bell Hooks' words: 'Until women accept the need for redistribution of wealth and resources in the United States and work towards the achievement of that end, there will be no bonding between women that transcends class'.[8]

Class is insidious in its complexity. The Furies, an early radical feminist collective in the United States, wrote that 'the small, indirect, and dishonest ways of behaving in polite society, are also ways of maintaining "the supremacy of the middle class—perpetuating the feelings of inadequacy of the working class"'.[9] For women, whose experience of interpersonal relationships is more intimate than men's, these interpersonal and often non-verbal impositions of class may be more salient than they are for men. Nancy Hartsock adds to this, 'confidence, verbal ability, ease about money, sense of group identity'—and is hopeful that feminism is developing new questions about these areas of class.[10]

Race and culture are further examples of categorizations which will lead women to experience oppression differentially. For example, white women cannot be blind to the fact that they have benefited from the oppression of non-white women.

Being oppressed by a dominant culture that defines a racial or cultural group as 'other' carries specific implications for women belonging to that group. For example, Anne Summers pointed out in her analysis of the sexist dichotomy informing Australian male culture, the Madonna and the Whore, that Aboriginal woman *has* to fall into the second category as the first automatically excludes her because of her colour.[11]

In discussing power and black women, black Australian Bobbi Sykes has written:

now we are talking about a matter of degree. White women merely have less power and control than white men. I do not doubt that some white women experience this state of lesser power acutely, but in comparison to both black women and black men, white women are extremely powerful and have control over a great deal of resources.[12]

She discussed the abortion debate in Australia as showing the differing racial interests for women. Where white women saw freer laws as a greater control for them over their own bodies, black women saw in them the threat of genocide.

Adrienne Rich has argued that race and class have been long-standing tools in the hands of misogynistic men, used to divide and rule women and to demand their allegiances to groups of men and not to other women. The privileges of being middle-class and white may seem considerable to those who are neither, but as Adrienne Rich comments 'whiteness did not preserve any woman from ownership by men' nor from brutal abuse and exploitation.[13]

Acknowledging these differences, women still share many common experiences across class and race boundaries. To undermine male power women need to form a cohesive revolutionary class. 'Sisterhood' is a moving and potentially radicalizing concept of united womanhood. It leads, for example, a Maori New Zealander and lesbian feminist, Ngahuia Te Awekotuku to say: 'I believe sexism to be the primary offence against humanity', because of men's brutality and violence to all women regardless of race.[14]

Hester Eisenstein claims that the concept of sisterhood is a 'false universalism'.[15] But socialist feminist Michèle Barrett has argued that it has value as both a term and a concept. She writes:

The accusation that the Women's Movement is 'middle class' in fact robs it of a justified recognition of the unique achievements in forging common objectives across the boundaries of class. The Movement is by no means restricted to women of one class. Although class divisions may cause problems that need to be worked on internally, the concept of sisterhood does have some political reality within the Movement.[16]

And women do share similar needs, for example, the control of their own fertility and their own bodies. They need to have access to free abortion, to have child care available twenty-four hours a day and to be able to walk the streets and live in their homes without experiencing violence, which knows no class. Discussing this point, Zillah Eisenstein comments: 'We will not know what our real class differences are until we deal with what our real likenesses are as women ... although our differences divide us, our likeness cuts through to somewhat redefine these conflicts.'[17]

It is important, as Bell Hooks has written, that women know that 'we do not need to eradicate difference to feel solidarity'.[18] There is a political strength in this feeling of solidarity and oneness, and in collective action which comes from it. This does not mean that every woman in every culture or class will identify more closely with women than with men, but that their relationships both with men and with other women have something in common. 'Many women are not aware of identifying with other women. They experience their lives as focused around their families and the men in their lives, yet share their feelings about those lives, and look for solace and support with each other'.[19] So Adrienne Rich has stressed the need to connect with each other to understand this common experience. She writes: 'To become truly educated and self-aware, against the current of patriarchal education, a woman must be able to discover and explore her root connection with *all* women'.[20] Work, for example, is not the same experience for a woman who is an assembly-line worker as for a woman who is a doctor or a house worker; but the importance of paid work to her self-identity is shared. Within these limits, and acknowledging them, we can attempt, as does Catharine MacKinnon, to 'include all women in the term "woman" in some ways, without violating the particularity of any woman's experience'.[21]

Dale Spender expresses the desire for both an acceptance of differences and a belief in women's shared oppression when she writes:

As a white westerner I know that I use more than my share of the world's resources, that I am privileged, that there are millions who are starving or being brutalised. I know that I profit directly from racism, war, exploitation, and that I cannot abdicate from that responsibility . . . I am part of that society. I cannot walk away: my life cannot be laundered . . . I am aware that I stand in relation to many black and third world women as men stand in relation to me . . . [but] we are all in the end subordinate to men; we share our oppression, our relative poverty (for according to the UN statistics women own less than 1% of the world's resources), our availability, our biological capacity. We are all 'other'.[22]

The problem that has a name—men

Men as a group enjoy the privileges of patriarchal power; the reality, and the sense, of being in control which they experience individually and collectively. A man exerts power over *some* men, but over *all* women. Men benefit from women's unpaid labour and emotional and physical servicing.

This is not to say that some men cannot betray patriarchy, supporting the values of feminism, just as white South Africans can be traitors to racism. Individual men can relinquish the power they have to abuse, deny, negate and otherwise control women, purely by right of their sex. As Rebecca Albury points out:

'Male control' doesn't essentially mean control by individual men. It means control which benefits men more than women most of the time. Far from each man exercising personal authority, things are much more complex. We live in a network of power relations that both defines 'masculinity' and ensures the success of individuals and activities that reinforce that definition.[23]

Coote and Campbell suggest that pro-feminist men have first to acknowledge that there is a conflict of interests between themselves and women. They need to 'put their own relationship to masculinity and patriarchal power into crisis if they are really to embrace feminism. By denying that their relation to women is contradictory, they promote themselves and their interests as universal. And they overlook the fact that they, men, are a problem for women'.[24]

Men are astonishingly ignorant about the forms of oppression that their own social group has practised on women and are often ready, having 'given the matter [their] acute political attention for a few hours', to expound their opinion on this—'what matter that thousands of women have worked on a problem for years'.[25]

Leonard is severe on masquerading 'feminist' men who she says claim feminism as their own for one reason—'exonerating men'. So they try to refocus women's attack away from patriarchy. The 'real' enemy, they say, is class, the stereotyping of men, or anti-feminist women. They speak of anti-male attitudes, 'as if hostility to the oppressor is the same as hostility to the oppressed'.[26]

In patriarchy, men cannot *not* have advantages, but they can choose not to exercise them. As Dale Spender has written:

Obviously some men begin to understand the ramifications of male power in the same way as I begin to understand the ramifications of white dominance; they listen, they learn, they learn to 'follow' the maps of others rather than insisting on taking their route. But just as there are no prizes for whites who act reasonably, neither are there pats on the back for men who choose *not* to use all the resources at their command against women.[27]

Men who choose to change themselves within the limitations of their lives, need to face and accept their accountability as a member of an oppressive social group. It needs the kind of emotional work unfamiliar to many of them.

Included as oppressors are those men who acquiesce in patriarchy, who do nothing to change their unfair advantage, who refuse to listen to women's arguments, who fail to empathize with the oppressed and therefore perpetuate the system. They may reinforce patriarchy by blocking potential change, by keeping women isolated on an individual level, prisoners to personal oppression; and by ensuring that the structures of society continue to do so at a group level. They also include the Fairytale Brigade, those men who offer marvellous visions of support which are never backed up and whose real aim is to manipulate and control while appearing to be a friendly ally.[28] And then there are the men who overtly and explicitly oppress women: the most visible enemy.

No man can experience the reality of women's oppression. Men can understand, empathize and agree with the analysis. But they cannot *be* female, with a history of being the 'other' and the daily disparagement and threat which that entails. As Catherine MacKinnon comments: 'Feminism is the first theory to emerge from those whose interests it affirms'.[29] Men can contribute by taking responsibility for sexism, refusing to collaborate with patriarchy, and actively fighting it, in their own place.

Patriarchal society is not just passively pro-male, it is actively anti-female, misogynistic and woman-hating. Ironically when women challenge patriarchy, they are labelled man-haters, anti-male, 'ball-busters', or castrating bitches. Of course the worst thing that can be said of anyone living in a patriarchal society is that they can't (or won't) get on with men!

Periodically, on a group or individual level, men reinforce woman-hating through force or the threat of it. Patriarchal structures or ideologies which maintain that women as a social group are worth less than men, and institutionalize inequality on those grounds, are also misogynistic. Discussing misogyny, Sheila Ruth writes that 'veiled by chivalry' it is a potent force in relationships between women and men and underlies 'not only rape, invective, and abuse, but also beauty contests, work segregation, menstrual taboos, mother-in-law jokes, old bag myths and themes, patronising etiquette, and current sexual mores'.[30]

Past and current history are sprinkled with examples of the institutionalization of misogyny. The burning of witches is notable. The witch-burning period at its worst covered the three hundred years from 1600 to 1900. Matilda Joslyn Gage estimates that nine million witches were killed for this crime after 1484, most of whom were not just burned. They were stripped, raped and then tortured for anything from forty-eight hours to a week—using boiling oil, thumb screws burning coals and other implements. A quote from Gage will indicate the numbers in which women were tortured and burned:

The Parliament of Toulouse burned four hundred witches at one time. Four hundred women at one hour on the public square, dying the horrid death of fire for a crime which never existed save in the imagination of those persecutors ... Remy, Judge of Nancy, acknowledged to having burnt eight hundred in sixteen years: at the rate of half a hundred a year. Many women were driven to suicide in fear of the torture in store for them. In 1595 sixteen of those accused by Remy, destroyed themselves rather than fall into his terrible hands. Six hundred were burnt in one small bishopric in one year; nine hundred during the same period in another. Seven thousand lost their lives in Trenes; a thousand in the province of Como, in Italy, in a single year; five hundred were executed at Geneva, in a single month.[31]

These were often women of superior knowledge who had healing gifts, skills which were to be professionalized in the form of a male-owned medical profession. They were women who lived alone, and therefore had no man to defend them, and threatened the established interpersonal structures, or who were old, and as Gage indicates, no longer sensual and therefore useful to men. The 'sin' of witchcraft was also useful for men who wanted to rid themselves of wives and could do so merely by accusation. After all, who knows a wife better than her husband?

The role played by the church in this enterprise is well recorded. The trials filled the coffers of the church because it gained the property of the witch and her family. Thus, as Gage shows, many wealthy women were burned, and the male separatist clergy fattened upon the torture and ashes of women.

Gage's book, *Woman, Church and the State* (1893), also documents the atrocities involved in the traffic of women, in the history of the institution of marriage, and the enslavement of women. Mary Daly's book *Gyn/Ecology* also covers acts such as Indian suttee (widow-burning), Chinese foot-binding, African genital mutilation, witch-burning and American gynaecology. Daly terms these 'sado-rituals', initiated by men as a punishment for women with the aim of controlling them.[32]

Kathleen Barry has documented the continuing international sexual slavery trade, with white women being beaten, tortured and seduced into sexual slavery in Eastern countries and Asian women being enslaved in Western countries. She names the ideology which accepts this situation 'cultural sadism', and appends the 1974 report of Interpol, which concludes that 'the disguised traffic in women still exists all over the world'.[33]

Christine Delphy analyses this battle as between two classes: social women and social men. She attacks Marxist feminists who refuse to so name the confrontation, writing that their refusal to consider women as a class and to consider men as the antagonistic class relates back in the end to its 'unthinkability'.[34]

Some women claim never to have experienced oppression, concluding that it does not exist, or can be easily overcome by individual effort. Again the personal is not in itself enough for social analysis. In order to bring about change and continue the revolutionary struggle, women need to understand that their own experiences may be different from those of many other women in the world. Speaking of the 1960s consciousness-raising experience, Bell Hooks writes:

Many women who had not fully examined their situation never developed a sophisticated understanding of their political reality and its relationship to that of women as a collective group. They were encouraged to focus on giving voice to personal experience. Like revolutionaries working to change the lot of colonised people globally, it is necessary for feminist activists to stress that the ability to see and describe one's own reality is a significant step in the long process of self-recovery; but it is only a beginning.[35]

However, it is threatening and unsettling to many people to step outside their own world experience or social class. As Peggy Seeger said: 'I was lacking gut experience and I had to arrive at my aware-ness of most women's problems, through observing, thinking about and listening to other women . . . I discovered how other women lived. It was quite a shock!'[36]

Texts such as those by Gage, Daly and Barry are important be-cause they remind women of the brutality of patriarchal control; of the existence of woman-hating; of the battle which does exist between the two social groups, women and men.

Women's identity under patriarchy

The 'self' usually includes all we can know and all we can feel about ourselves. For example, Jersild writes:

A person's self is the sum total of all he [she] can call his [hers]. The self includes among other things, a system of ideas, attitudes, values and com-mitments. The self is a person's total subjective environment; it is the distinc-tive centre of experience and significance. The self constitutes a person's inner world as distinguished from the outer world consisting of all other people and things.[37]

R. D. Laing has written that 'one's self identity is the story one tells one's self of who one is'. But what has this meant for women in a male-dominated culture where the power to name is in men's hands?

Men have described themselves as the norm and everything non-male as 'not-normal' and 'other'. Studies of attitudes of psychologists to healthy males, females and adults reinforce this bias. Their descrip-tions aligned healthy adult with healthy male, but healthy female was not defined as the same as a healthy adult.[38]

The elements of women's self-definition are said to be very different to those of men's identity, and have been ascribed an inferior and negative value. For example, sexually women can be nymphomaniacs, sluts or frigid, but never virile. Furthermore, the personality char-acteristics allocated to women are those which are primarily socially undesirable. Judith Bardwick and Elizabeth Douvan give examples of these. The traditional characteristics are:

dependence, passivity, fragility, low pain tolerance, non-aggression, non-competitiveness, inner orientation, inter-personal orientation, empathy, sensi-tivity, nurturance, subjectivity, intuitiveness, yieldingness, receptivity, inability to risk, emotional liability, supportiveness.

The corresponding male characteristics are:

independence, aggression, competitiveness, leadership, task orientation, out-ward orientation, assertion, self discipline, stoicism, activity, objectivity, analytic-mindedness, courage, unsentimentality, rationality, confidence, and emotional control.[39]

The common term 'opposite sex' takes on its real meaning when we look at these descriptions. Society values most highly the character-istics ascribed to men,[40] and the characteristics which are positively valued in women are those which do not threaten male power and control, which are best suited to gratifying his needs, and which enable society to relegate woman to the domestic sphere on the grounds that her personality suits her best for this role.

The role of woman as nurturer is rigidly prescribed. Submission to men (husband or father) and subjugation of her individuality are essential. Self-sacrifice in the precise meaning of the word is lauded. Within these terms, women are expected to realize themselves by fostering fulfilment only in others, which leads to self-alienation. They are expected to fulfil themselves by proxy. Ironically, while this self-sacrifice is demanded of them, women are disparaged for giving it. Mary Daly has commented:

in contrast to male modes of self-sacrifice, which are rewarded with the ecstasy of merging, the self-sacrifice imposed as an ideal upon most women is the most radically unrewarding handing over of their identity and energy to individual males: fath' s, sons, husbands—and to ghostly institutional masters.[41]

Women have been struggling against this imposition of self-sacrifice or self-killing for centuries and those who succeed often pay a price for their independence Matilda Joslyn Gage said in 1893:

But woman is learning for herself that not self-sacrifice, but self-development, is her first duty in life, and this, not primarily for the sake of others but that she may become fully herself; a perfectly rounded being from every point of view.[42]

The inducement to yield self has been 'love'; love of men and children but not of self. Woman's love is expected to be selfless and therefore pure. This way women have been trapped into selflessness, dangerous both to mental health and to life satisfaction. The concept of romantic love has also been used to de-activate women's struggle against patriarchy. As Andrea Dworkin has written: 'Love is always crucial in effecting the allegiance of women'.[43]

Selfhood is created in interaction with others. But woman's sense of self-esteem has been made abnormally dependent on male approval. Woman is the 'object' to the male 'subject'. To move away from this dependence on men for selfhood is difficult. it is not encouraged by malestream society because it will call other areas of women's lives into question. But relationships with other people should foster self-definition and strength, not dependence.

The objectification of woman and her feminine identity is internalized as well as being externally imposed. MacKinnon comments on this objectificaiton when she discusses 'the pain, isolation, and thingification of women who have been pampered and pacified into non-personhood—"women grown ugly and dangerous for being nobody for so long"'.[44] Here she is discussing middle-class women and the reality of their deprivation while pointing out that those suffering materially might find it difficult to sympathize.

Carrying her analysis through sexuality into the area of identity,

MacKinnon interprets the characteristics of feminine identity as creating sexual access for men:

Vulnerability means the appearance/reality of easy sexual access; passivity means receptivity and disabled resistence, enforced by trained physical weakness. Softness means pregnability by something hard. Incompetence seeks help as vulnerability seeks shelter, inviting the embrace that becomes the invasion, trading exclusive access for protection ... Narcissism insures that woman identifies with that image of herself that man holds up ... Masochism means that pleasure in violation becomes her sensuality.[45]

Feminists are drawing attention to the concept of femininity and its part in the oppression of women. Susan Brownmiller's *Femininity* examines the 'ethic', its longevity historically, its origins and its endurance. She concludes that it is harmful to women, and results in 'self-imposed masochism (restraint, inhibition, self-denial ...)'.[46] So women are encouraged to collude in this process.

Jill Matthews has dealt with femininity in a different way, analysing the development of the 'good' woman in Australia during the last century. She does this by considering the case histories of women driven 'mad' by the imposition of the feminine ideal. Femininity, she concludes, is prescriptive; the meaning of it is determined for a woman by 'each individual and institution which has power and authority over her'. The definitions invariably contain contradictions and idealized qualities, ensuring that femininity is impossible to achieve. 'To be a woman is thus necessarily to carry a sense of failure'.[47] The feminine ideal robs women of 'any individuality, that is self-serving or self-created, not useful to the male in his scheme of things'.[48]

Self-esteem entails a confidence in one's personal sense of worth. If this is low, a woman is isolated and dependent at the same time: unable to act and take responsibility for her own life; and more importantly unable to enjoy a satisfaction with herself and her actions. Disparagement of women in our society lowers self-esteem, and encourages group self-loathing and self-hate so that women do not value other women and keep themselves apart from them. As Hélène Cixous has written:

Men have committed the greatest crime against women. Insidiously, violently, they have led them to hate women, to be their own enemies, to mobilise their immense strengths against themselves, to be executants of their virile needs. They have made for women an anti-narcissism! A narcissism which loves itself only to be loved for what women haven't got! They have constructed the infamous logic of anti-love.[49]

This diminution in self-esteem can lead to frustration, hostility, anxiety or depression, and self-esteem is lower in women than men because women are denied real identity. Benson summarizes the results of disparagement for women when he writes:

Thus the female ... often comes to perceive herself negatively no matter what happens. In a sense she cannot win. If she is too attractive, she may be used as an object. If she is not attractive enough, she may not be desirable at all. If she is intelligent, men may be afraid of her; if she is stupid, she will be treated like an article of furniture. If her sex drive is high, she is a tramp. If it is low, she is not a woman. It is no wonder, therefore, that women in our culture find it extremely difficult to develop their real potentialities without experiencing a great deal of emotional turmoil and stress.[50]

A person low in self-esteem is also more pliant, more open to moves which will, no matter how temporarily, raise her/his sense of worth. If the constant disparagement by a powerful individual or group of another less powerful individual or group leads the less powerful to accept a negative image of themselves, they are easier to control. This is done through the simple method of small doses of reward which make the less confident feel good about themselves. They then strive to please the powerful in order to be 'rewarded'.

How much healthier and more satisfying it would be for women to seek inner approval and satisfaction and to reject the dependence/punishment/reward cycle imposed by patriarchy. Dickstein clearly articulates the intimacy and warmth of self-knowing when she says:

One can never know, or be known by another self in any way which approximates the 'feelings of warmth and intimacy', the urgency or the poignancy, accompanying one's own self. The man [sic] who has attained self consciousness can choose, and develop and change his milieu—and if he loves that of which he has become aware he will strive to accomplish, he will defend himself if he must, and if he be buffeted by fate or his fellows, he can always retreat to his own sense of worth or finally his sense of pride.[51]

So should it be with women.

By being negated, de-selfed, dispossesed and silenced, women become manipulable. The collective in *Our Bodies, Ourselves* talked of feeling limited, 'dependent creatures with no identities of our own'.[52] The experience of not being real, or not existing, is often expressed by women as they struggle to hold on to a selfhood continually eroded in patriarchy and aggravated by enforced isolation from their sisters.

Discussing this, Andrea Dworkin points out the ways in which women's sufferings and the history of brutality towards women are constantly erased or silenced.

And if a woman, an individual woman multiplied by billions, does not believe in her own discrete existence and therefore cannot credit the authenticity of her own suffering, she is erased, cancelled out, and the meaning of her life, whatever it is, whatever it might have been is lost. This loss cannot be calculated or comprehended.[53]

When discussing the crucial importance of self and a sense of identity, William James pointed out the importance to a man of the 'recognition which he gets from his mates'. In a vignette of what James sees as the ultimate horror for the individual, unintentionally he writes of feelings which many *women* have had, and the desperation which drives them to accept the identity offered to them in order to exist physically and psychologically.

No more fiendish punishment could be devised, were such a thing physically possible, than that one should be turned loose in society and remain absolutely unnoticed by all the members thereof. If no-one turned around when we entered, answered when we spoke, or minded what we did, but if every person 'cut us dead' and acted as if we were non-existing things, a kind of rage and impotent despair would ere long well up in us, from which the cruelest bodily torture would be a relief; for these would make us feel that however bad might be our plight, we had not sunk to such a depth as to be unworthy of attention at all.[54]

Women are also entrapped by their belief that only the 'odd' man is cruel or brutal or unjust. They continually rely on men to be fair and continue to be surprised when they are not. They experience a faith in the powerful which is a necessary condition and a result of being powerless. Wrong describes the experience thus:

Helpless to resist coercion, fearful of punishment, dependent on the powerful for satisfaction of basic needs and for any opportunities for autonomous choice and activity, the powerless are inescapably subject to a will to believe in the ultimate benevolence of the power holder, in his acceptance in the last analysis of some limits to what he will demand of or inflict upon them, grounded in at least a residual concern for their interests.[55]

One price paid for this compromise is having the self split in two, so that a woman's sense of who she is comes to be supplanted 'by a sense of being appreciated as herself by another'. Thus 'men look at women. Women watch themselves being looked at...the surveyor of woman in herself is male: the surveyed, female'.[56] Woman's self has become objectified even to herself, and to be real, to exist, she now has to comply. She learns a 'grateful complicity in exchange for survival and self-loathing to the point of extinction of self'.[57] Women have thus internalized patriarchal definitions of 'woman'.

Only respect for the self makes resistance conceivable. And resistance to annihilation *has* constantly existed. The survival of women and their strength is evidence of it. But the attack on woman's core self continues and is important. As Mary Daly writes: 'The primordial universal object of attack in all phallocentric wars is the self in every woman'.[58]

But this form of self-constraint through patriarchal definitions can be changed, and the women's movement has been strong in its advocacy of self-reassessment for women. Anne Koedt comments that the oppression of women is a dynamic that can be understood 'in terms previously called "non-political"—namely the politics of the ego', which she defines as 'the sense of the individual self as distinct from others'.[59] Similarly, Nancy Hartsock stresses the need to begin by understanding our own lives and experiences before placing them in the context of wider social institutions. She comments that 'a fundamental re-definition of self is an integral part of action for political change'.[60]

Politically, consciousness-raising groups have played a large part in bringing women in touch with other women; and through a discussion of their common needs and feelings, in touch with themselves. Glenys Huws says the experience is similar to that of male initiation rites, culminating in a transition to a new personal and social identity; a 'birthing of the new self'. Like all formative experiences in adulthood which mean a break with the internalized view one has of one's self from childhood, this is an emotional experience, bringing feelings ranging from anger to exhilaration. It also involves what Huws calls a 'rupture with the self-understanding of the past, in the struggle to affirm the new self'.[61]

But before women can create new self-identities, they need to understand how woman has been defined by men; and to analyse the processes which create that definition.

Self-identity: process and content

Miriam Dixson in her analysis of Australian women in *The Real Matilda* defines self-identity using a quote from Erikson which stresses process. Self-identity is a 'process "located" in the core of the individual and yet also in the core of his [her] communal culture . . . [it] depends on the support which the young individual receives from the collective sense of identity characterising the social groups significant to him [sic]: his class, his nation, his culture'.[62] We should also add 'sex' to this list.

The processes of patriarchal society through which the self takes shape have been outlined briefly, for example the socialization process which begins at or before birth and continues throughout the adult years. This process imposes norms of behaviour, as for example when a girl is told 'not to be a tomboy'. Little girls are *taught* not to climb trees, to wear dresses, to sit with legs femininely crossed, to smile and beguile to get their way. Boys are *taught* not to cry, not to show vulnerability, to assume rights of space, to be direct and take control. So, through sex difference (biological), gender difference (cultural) is created. Women are rewarded and punished into assuming the 'right identity'.

They also learn from personal experience. But this experience often conflicts with the 'rules' of being 'woman'. Women's experiences have been continually invalidated and treated as unimportant or not real by those who have the power to create definitions of reality. Feminism has stressed the validity of personal experience. The dichotomy between the personal and public spheres has been deliberately constructed, to exile women to the private domain. By pointing out the interrelationships and overlapping experiences in the personal and public spheres, feminism has attempted to break down this artificial construction.

The self is both consistent and variable—to a degree it is situational—

moulded and influenced by the experiences of a particular historical time, and by its own personality and character. It is also controlled and limited by social institutions and structures which constrain the availability of some experiences for the individual.

The ideology of male right permeates social practice and forms a part of the social unconscious which passes on to the next generation. Ideology formation and development of culture include the control of knowledge and naming. Language itself has been shown to be a male-constructed set of signs.[63] Men have acquired and control the power of naming what is valued and powerful—for example, they name only paid labour as valuable and by defining woman's place in the family, entrap her unnecessarily in child care, and force her into economic dependence.

Ideology reinforces structures which are designed to benefit men, for example the economic, political, legal and religious structures. Through these, women's subordinate position and lack of rights can be institutionalized. In Australia, for example, women were defined out of the workforce and into the role of 'dependent' in 1891 by a definition of 'occupation' which centred on the receipt of an income, not the notion of productivity. In 1907 the Harvester Case developed the notion of the 'basic' or 'family' wage, based on the belief that any working woman is awaiting marriage or has (should have) a bread-winning male as supporter. Jobs were classified as men's or women's work and men paid a higher rate. So lower wages for women were institutionalized, justified by the ideology of the 'family'.[64] Once such patriarchal (and hierarchical) structures are in place and difficult to dismantle, men can relinquish any personal responsibility for women's oppression, by pointing to the impossiblity of changing the 'system'.

These processes operate constantly to control women—their activities, their lives, their choices, their definition and valuing of self—though the degree of oppression or control varies. For example, the law has been changed to be more favourable (though still not just) towards women, and education is more readily available than it was in the nineteenth century.[65] But discrimination still exists in pay—women do not get equal pay for equal work—and in the sex segregation of jobs.

There is not space here to explore the 'processes' in more detail, though they will be clarified further through the analysis of self in the following chapters. This book will concentrate mainly on the content of self for women. However, process and content overlap and cannot be distinctly differentiated. For example, a discussion of women's work will show how women's labour in its various forms is devalued, unpaid or underpaid. This leads some women to perceive themselves as 'not working'. This perception is reinforced ideologically, but also by the structures of the gendered workforce. And women's emotional dependence is both a definition of woman *and* a process by which women are disempowered or de-selfed. So the definition of self and the means of its construction are intertwined.

The content of self has several elements. It is dynamic, containing several interconnected dimensions, physcial and non-physical.[66] Firstly, there is the material or physical self, which includes the 'shell' in which the self resides and the image a person has of the body. This body-image often differs from reality, demonstrating that the experience of the body is partially a construct. So the anorexia nervosa victim experiences her body as she perceives it to be—large and fat rather than emaciated. Schilder has commented that body-image also includes 'whatever originates out of the body',[67] for example, voice, semen, or menstrual blood, and this has implications for the ways both women and men relate to women's bodies. The body is one of the dimensions of self and the individual's sense of body-image has a bearing on the way she will act. In Chapter 1 I will consider the way woman is reduced to body only within patriarchy.

The self is also composed of the mind and the intellect. These are often presented in a dichotomous fashion as opposites to the body, and to feeling or imagination. Man has split the emotions from reason and rational thought, associating the former with women and thus devaluing them, particularly in decision-making. But Joan Cocks points out that reason and emotion are not antagonistic to each other.

Because emotions are felt does not mean that there is no reason in them . . . just as the intensity of emotion does not indicate that it is irrational, the coolness with which an argument is made, while it may show self control on the speaker's part, does not mean that the argument is innocent of special passions.[68]

Both the mind and the imagination are part of the development of culture. Dealing with the dichotomy established by patriarchy, I will consider in Chapters 2 and 3 the male construction of science as a rational system and its relation to women, as well as the representation of women in culture through an examination of writing and its construction as a male enterprise. Through writing, women are currently redefining selfhood and expressing resistance.

Work, its value to others, and a sense of purpose from it, are also important to a sense of identity. It puts a value on a person's labour and time. Work that is socially valued creates a sense of self-esteem for the individual, who has a place in the world which is vital and necessary. In a society which values money, paid labour is more highly valued than unpaid work. In Chapter 4, I will consider the work that women are deemed to be doing by men, and show how work is constructed to devalue women so they are underpaid, or their work is not recognized at all.

Self is also created from community with others and from the relationships formed with them. These reflect the self back to a person and help to modify or reinforce behaviour.[69] The social self becomes the individual, and it is created in relationship with others. Chapter 5 will deal with the way relationships have been constructed so that woman is defined as solely for man, the nature of that relationship, and the way women have constantly struggled against it in efforts to regain their woman-identified and self-identified selves.

The final chapter is a conclusion dealing with power. Power is the element linking the constructions of woman's self together in order to constrain and control her. The process of the imposition of male power is analysed.

Male knowledge-construction has made women's history invisible, causing them to become divorced from their own past and their own consciousness of the world. The validation of ideas and thoughts of individual women comes from knowing that others have had similar experiences or ideas. Without that chain of women through history who speak as they do, women have no sense of the tradition of shared experience and many feel isolated and alone. A sense of being part of a strong and continuing group of people who express similar opinions and have similar experiences is a powerful part of a person's identity, and can form a strong and cohesive power base which enables groups to push for change. As Leghorn and Parker write:

> Recognition by less powerful individuals of their common experiences and of real differences from the dominant culture has been a tremendous source of self-knowledge and thus self-respect. To be grounded in a strong sense of self, and to be self-identified . . . have served as a solid base from which to act in one's own behalf . . . [70]

Concomitant with male power has been women's resistance. That continuing struggle against patriarchal definitions of woman, is re-affirmed throughout this book. Historically there are a plethora of quotable instances, and no one less vehement than the others, in which women discussed, debated, exemplified, satirized, or ridiculed the injustice which is handed out to them. For example, Mary Astell (1638–1731) in England as early as 1694, in a *Serious Proposal to the Ladies*, wrote that women were denied education so that man could seem to be superior; that man created the stereotypes imposed on women and then scorned them for it:

> Women are from their very infancy debarred those advantages with the want of which they are afterwards reproached [and are] nursed up in those vices which will hereafter be upbraided to them.[71]

American Mariana W. Johnson, in the Memorial to the Constitutional Convention adopted by the Women's Rights Convention in Sale, Ohio, in 1850, wrote:

> We believe the whole theory of the common law in relation to women is unjust and degrading, tending to reduce her to a level with the slave, depriving her of political existence, and forming a positive exception to the great doctrine of equality as set forth in the Declaration of Independence . . . At the marriage altar, the law divests her of all distinct individuality . . . Woman by being thus subject to the control, and dependent on the will of man, loses her self-dependence.[72]

In Australia in 1905 in the *Women's Voice*, a journal of the Women's Reform League, one woman wrote:

The age of consent legislation seems to have been instituted in the interests of sensuality, for the interpretation of the law makes it protective of man rather than of woman. It implies that above the statutory age, consent is taken for granted unless her resistance be almost to the death. If the aggressor can prove the voluntary acquiescence of his victim, no matter how or when it was obtained, he need not make any attempt to deny the commission of the crime, for it is only held to be complete when it is committed by force, without the consent and against the will of the woman. In a court of law it was decided that a girl must resist until exhausted or over-powered, for a jury to find it is against her wish—that resistance and dissent ought to have been continued to the last. No wonder that under this interpretation of the law by the judge and jury, the popular sentiment is that 'at the age of consent' means that the State consents to this criminal injustice—consents to legalize the destruction of virtue and protect the man at the expense of the woman.[73]

If women are acted upon and constrained and constructed, can they then act to change that situation? Feminism has shown that women's resistance can be effective in defying patriarchal definitions of woman and altering women's oppression. As Clare Burton has said: 'The feminist movement is a visible manifestation of the fact that women are not simply passive recipients of a dominant ideology which imposes on them a definition of their place.[74]

While women are products of their history and social construction, they can also shape their lives within those constraints, and sometimes push the boundaries aside. Self is not fixed, but is formed early and to change it requires struggle. It is the product of the dialectic between the forces which impose upon and create the self, and the individual's attempts to interact with and recreate her social existence.

In 1829 Fanny Wright upbraided women for accepting their subordination. She urged women to begin with themselves first, 'to analyse what we know, how we know it and whether we should begin to know very different things'. She challenged: 'Go, then! And remove the evil first from the minds of women, then from their condition, and then from your laws.'[75] Her words are still necessary in the 1980s.

1

Sex, Biology and Self

From early societies to the present day, woman's physical self has preoccupied man. In the body lies the most obvious sex difference, and around it have been woven images, myths and taboos. Woman's body has been both reified and degraded: has inspired both awe and fear. For woman too, her body is the first reality, the physicality in which the self resides. It has always been a battleground where the power play of the sexes takes place. Brettana Smyth, a widow who sold contraceptives and wrote books and pamphlets on contraception and women's control of their bodies, wrote in 1913 that 'it is to the body one looks for the Magna Charta of feminism'.[1]

Men have created the image of woman's body to which she should aspire, and have defined the purposes for which this body should be used. Through biological difference they have attempted to define the 'natural' behaviour of women, using science as a primary tool.

Biological difference and the 'natural'

In essence, little of the ideological base of the debate has changed since the days when woman's menstruation was used to define her as mysterious, strange, unstable, and, in her connection with nature, both uncontrollable and dangerous. Evidence for the strength of either biology or culture as forming the differences between women and men focuses on evidence from physiological differences, especially with respect to hormones and the brain, gender differences and behaviour, socialization arguments and anthropological debates.

The debate over 'natural' behaviour has been called either the nature/nurture debate, stressing analysis through consideration of biological as against learning/socialization processes and occurring, for example, in the discipline of psychology; or the nature/culture debate (in, for example, anthropology), which emphasizes biology versus a wider social structure analysis. Politically, if woman's position can be proved to have come about because her nature has determined it, her position is immutable, unchangeable. But if culture or nurture determines her position, change is possible in patriarchal structures and ideology, and within the woman herself.

Men have argued that women and men are innately different. This came to be dichotomized as superior/inferior, that is, women are different to men, the norm, and are therefore peculiar. Notably, woman has always been the site of the study. So the question has been how is it that women are different, rather than why it is that men are so odd! The stress has fallen on differences rather than similarities in behaviour; on man as the norm; and on woman as the 'problem'. This has been challenged by women, particularly since the post-1960s wave of the women's movement, which has pointed out the bias within research, showing language, methodology and theory to be androcentric (male-centred) or phallocentric.

Biological and physiological differences between the sexes exist on a number of levels. Women and men look different physically, with respect to body size, weight, musculature and hairiness for example. But individual differences play a large part: for example, even though men on the whole are stronger and women have more body fat, individual women can be stronger than some men, and some men fatter than some women.

Chromosomally, the sexes differ. Of the twenty-three pairs of chromosomes which determine human inheritance, *one pair* controls sexual development. For females, the two chromosomes are designated XX, and for males, XY, the Y chromosome carrying 5 per cent *less* genetic material. Interestingly, genital differences only appear in the embryo in the seventh week of development, and then only if androgens are generated to *suppress* the growth of ovaries. So all human life *begins* by being female and if not interfered with, the 'natural' product will be woman.[2] Montagu has called this 'the natural superiority of women'.[3] It could be argued then that men are not the 'norm' at all. Valerie Solanis wrote:

The male is the biological accident: the Y (male) gene is an incomplete X (female) gene, that is, has an incomplete set of chromosomes. In other words, the male is an incomplete female, a walking abortion, aborted at the gene stage.[4]

Adler, discussing the impact of the information that all life begins as female, wrote;

It was ... a shock equivalent to the theological declaration that God was a woman when embryologists discovered that the basic pattern of the embryo was neither male nor undifferentiated nor, as Freud believed, bisexual, but female in structure ... It would be like saying that God created Eve first, then added to her body an androgen-soaked rib to produce a belated Adam.[5]

In terms of identity, this information should increase women's positive view of themselves as inherently normal and not 'the other'. In fact, man is 'the other' biologically. It could be argued then that the clitoris is not a substitute penis, rather the penis is an 'exaggerated clitoris'. Furthermore, the male is the sex which passes on the majority of sex-linked diseases (haemophilia, for instance); more male foetuses than female are naturally aborted; and the male lives for a shorter period than the female.[6]

Hormonally the sexes also differ. The major female hormones are oestrogens and progesterones produced by the ovaries, while the primary hormone which develops and maintains maleness is testosterone, one of a group of androgens. However, *both* sexes produce small levels of testosterone and oestrogen respectively. Great stress has been placed on the influence of these differing hormones, as they are said to create male and female behaviour.

Aggression is a good case study here in men's justification of their violent and often anti-woman behaviour. Aggression is anti-social, violent and leads to the infliction of pain and harm on another. Yet it is seen as making men strong, powerful and dominant, and is institutionalized in war and sport. In the process it has come to be seen as a desirable masculine characteristic and to be innately part of man's character. Alan Alda, commenting on this, writes:

Until now it has been thought that the level of testosterone in a man is normal simply because they have it but if you consider how abnormal their behaviour is, then you are led to the hypothesis that almost all men are suffering from testosterone *poisoning*. Testosterone poisoning is particularly *cruel* because its sufferers usually don't know they have it. In fact, when they are most under its sway they believe that they are at their healthiest and most attractive. They even give each other medals for exhibiting the most advanced symptoms of the illness.[7]

But the findings on links between testosterone and aggression have been contradictory. Studies on rats have shown that injections of the hormone will produce fighting behaviour. However, studies which involve injections of oestrogen into female rats have also been found to increase fighting behaviour in adult females.[8] It may also be inappropriate to link male human behaviour with that of rats and then to generalize about sexual differentiation in humans.

Some studies of monkeys have found that the environment can influence testosterone levels. Male monkeys in an all-female colony were found to have high levels, but when they were relegated to low status in an all-male colony their testosterone levels dropped significantly. Similarly, some studies on men show that stress reduces the testosterone level.[9] The level of testosterone can thus be changed by external or social conditions.

Lips and Colwill suggest that it is possible that physiological differences between the sexes only create a 'preparedness to learn aggression' for males.[10] It is more likely that social conditioning of males rewards aggressive behaviour, to ensure the dominance of men as a social group. Anthropological evidence cited by Parker and Parker indicates that cross-culturally there is a consistency in male aggression.[11] It appears to be a predominantly male trait, and one which ensures that men continue to be dominant because of the ultimate weapon of physical violence.

Aggression, testosterone and 'sexuality' in men have often been strung together. Groth, Burgess and Holmstrum examined 133 convicted rapists and ninety-five raped women. Sexual need was the excuse used by rapists, but the researchers delineated two basic motivations for rape associated with dominance and aggression: the aim of 'power or conquest', and the aim of 'violent retaliation for perceived wrongs and rejection by women'.[12] These kinds of studies indicate that rape is unrelated to a biological urge caused by testosterone in men, but is an act of aggression intended to release anger and exert power over women.[13]

Male dominance and women's subordination through biological difference are justified in order to maintain the oppression of women. Even if one accepts some biological difference, one is still left with the question, 'what are the innate or immutable *behavioural* characteristics of the adult that can be attributed to the hormonal differences in the two sexes?'.[14]

One attempt to answer this question has arisen in studies of humans whose chromosomal development was not 'normally' male or female. For example, the Money and Ehrhardt studies on hermaphrodites who were male chromosomally, but female or 'indeterminate' in their anatomy, indicate that biology is important but that the 'sex of rearing takes precedence over the chromosomal sex'.[15]

Other studies have considered neonates, or newborn children. The reasoning here is that gender differences exhibited in the newborn must be innate, as the effects of socialization could not have had time to influence behaviour. Problems with this work include the fact that behaviour in the newborn is fairly limited. Williams discusses some of these studies and their methodological problems and concludes that no results on gender differences and behaviour clearly attribute specific behaviours to a specific sex.[16]

Also problematic is the research conducted with animals. These studies are aimed at showing the naturalness of sex-stereotyped behaviours based on the female's child-bearing role. The work done with baboons, for example, has been used to indicate the naturalness of male aggression because baboons are strikingly sex-typed. Male rhesus monkeys are similarly patriarchal, while the females are calm and gentle. The Harlow monkey studies are the best-known studies and were originally designed to consider mothering. Harlow isolated baby rhesus monkeys and reared them with surrogate mothers which were inanimate wire figures shaped to look like a monkey. The monkeys behaved towards the mother according to their appropriate sex, with males aggressive and females passive, leading Harlow to conclude that 'we had demonstrated biologically-determined sex differences in infants' behaviour'.[17]

Studies of animal groups which do not support the equation that woman equals child-bearer and child-rearer, which means 'naturally' less powerful and suited only for the domestic, are studiously ignored. Lancaster has reported that in long-term studies, animal groups do not cluster around the males. She argues that most primate groups are mother-centred and males attach themselves to one or other of the female-centred groups.[18] Other researchers have found monkey groups in which the fathers take control of the care of the young, leading to the conclusion that in these groups 'the major division of labour is not male/female but rather child rearing/non-child rearing. It is the function of caring for a child that limits the extent of other functions, not the sex of the animal doing the caring'.[19]

Results from the animal studies are sifted through the lens of the researcher, usually male, who lives in a patriarchal milieu and whose belief-systems lead him to perceive and interpret the world the way he wants to. Why are certain animal groups studied in preference to others? Baboon behaviour *is* hierarchical and sex-stereotyped. Yet chimpanzees have 'very fluid relationships with one another that are difficult to stereotype by sex except for the fact that females nourish the unweaned young'.[20] These, however, have been rarely studied. They clearly would not suit the patriarchal justification of women's position, as they do not provide evidence in support of male dominance and male superiority over females. And when aggression in male baboons is found, why is it presented as evidence in support of male *superiority*, stressing the value of aggression over gentleness? So-called 'natural behaviour' has been idealized by men and used to present themselves in a favourable, more powerful light. It has been used to show that male dominance is necessary and normal; that to change the balance would be unnatural.

But nature is only seen to be 'right' when it suits patriarchy. There is evidence within anthropology that males within a variety of monkey, baboon, and ape groups commit infanticide. This is particularly true when a new male enters a breeding system. Hrdy documents the observed instances in detail and though she comments that it is not a 'common' occurrence, the frequency is astonishing. She writes that:

powerful intellectual traditions in the social sciences hold that primate societies are integrated structures in which every animal co-operates for the good of the group as a whole. Nowhere in this scheme is there room for such unabashedly selfish behaviour as the killing of another male's offspring by a rival progenitor, to promote his own reproductive success.[21]

Hrdy notes that over and over again the evidence has been ignored or treated as folk lore because anthropologists did not want to believe it. The concept of males being 'natural' infant killers has not been stressed alongside their role as the culture-makers.

Culture and the 'natural'

The analogies between animal behaviour and human behaviour are dubious, primarily because they do not take into account the complexity of human social structures and behaviour, and the influence of language and consciousness. The tools of power which enable humans to manipulate each other have irrevocably changed whatever may have been 'natural' behaviour for them.

Cross-cultural studies have also been dragooned into attempts to tie women's behaviour to nature. If, by studying a range of cultures over many differing periods, it can be shown that men have always ruled and women have always been subordinate, patriarchal men can argue that it must therefore be a 'natural' state of affairs and should continue.

Within anthropology, past literature has attempted to show that in all cultures 'male' has equalled 'culture' while 'female' equalled 'nature'. This runs parallel to the nature/nurture argument and attempts to justify the connection of women with the domestic sphere and their exclusion from public power by virtue of child-bearing and -rearing. Women's bodies and their life-creating ability are used as justification for divisions of labour and power.

Difficulties within the cross-cultural literature lie in trying to ascertain the 'true' relations between the sexes, which have been studied in the past from an androcentric or patriarchal viewpoint.[22] It is difficult also to ascertain the status of women and men in various societies and the differences in power they wield. For example, Aboriginal society has often been studied under the Western assumption that the men were the important people to talk to, thus erasing the knowledge of women and their part in the workings of Aboriginal culture.[23]

Furthermore, once the ascribed power of women is ascertained, it is difficult to tell whether this power can be exercised to control or stop men from carrying out certain actions. The Iroquois Indians, for example, are well known for their more egalitarian role distribution. Women had the power to select the chief and religious leaders (though they could not be selected themselves) and to reject lazy males from the long houses, which they solely controlled. They owned the land and controlled the food, which represented wealth, and without which men could not go on a hunt.[24] However, Parker and Parker comment that even in this more egalitarian situation, important aspects of status of Iroquois women derived from their links with male kin.

Though many studies are located in the past looking at early cultures, others consider current differences across cultures. Many of the former concentrate on hunter—gatherer societies, creating the image of the women as burdened with child-rearing, waiting by the campfire for the strong hunter to return with food. This enabled anthropologists to link man with culture, stating that males had needed to band (bond) together to hunt and therefore had to communicate. They were thus the creators of social skills and language, while the passive receptors, women, waited.

Slocum has attacked this vision. She states that it leads to the 'conclusion that the basic human adaptation was the desire of males to hunt and kill. This not only gives too much importance to aggression, which is after all only one factor of human life, but it derives culture from killing'.[25] Slocum sums up the patriarchal picture of hunter—gatherer society thus:

The females were more burdened with dependent infants and could not follow the rigorous hunt. Therefore they stayed at a 'home base', gathering what food they could, while the males developed cooperative hunting techniques, increased their communicative and organisational skills through hunting and brought the meat back to the dependent females and young. Incest prohibitions, marriage, and the family (so the story goes) grew out of the need to eliminate competition between males for females. A pattern developed of a male hunter becoming the main support of 'his' dependent females and young (in other words, the development of the nuclear family for no apparent reason).[26]

Men are also said to have created the first implements, these being tools and weapons for fighting.

But Slocum argues that even in current hunter—gatherer societies, it is clear that the primary food supply comes from the gathering by the women which is a reliable food source, unlike hunted meat. Women and children also hunt small animals. The dependence of small children would also have necessitated organization for feeding and skills in caring for the young. The passing on of knowledge and customs falls to those caretakers of the young, and *women* would thus have needed to develop symbolic communication and language. Their co-operation would draw them together into a social unit. Furthermore, the aids needed in gathering would include containers, digging sticks, and slings in which to carry the babies. These were thus among the first 'cultural inventions'. But evidence of their existence would not be as concrete as an axe, because bags or slings for example, would have been made of more perishable materials.

Estioko-Griffin and Griffin discuss the women of the Agta people in the Philippines. These women hunt game with machete and bows and arrows and contribute equally to the food supply. Only women with very young infants cease hunting for a short time. The authors argue that the Agta women have a substantial amount of equality in economic terms and in terms of their marriage partners. They comment that many Australian Aboriginal societies also have women hunters, though these do not on the whole hunt kangaroos. They point out the need to make the definition of hunting more flexible. What size animal does it include, and is net-hunting included?[27] Discussing Australian Aboriginal culture, Hiatt points out the practical implications of food availability: 'in hunting societies women are more important than men as food providers in areas where gatherable foods occur in abundance; men are more important than women as food providers in areas where gatherable foods are scarce.'[28]

In recent years, much more has been written and understood about the complex role of women in traditional Australian Aboriginal society. Intergroup differences abound. But Berndt, for example, has summarized some points about Aboriginal women: relative to men, they are domestically-centred or family-centred; there are no institutional channels through which they can change the system or their position in it; tied by myth and ritual to the system they show little indication of *wanting* to change it: and basically they are economically dependent on men, though have a limited role in ceremony and ritual (thus blunting the pressure to change).[29] Diane Bell, on the other hand, has found women to inhabit an essentially separate sphere which men recognize and respect, though their position has been eroded by land loss and white values. Aboriginal women may develop their own power base and she claims that 'we know too little of the female half of society to argue for male dominance as an enduring, timeless reality'.[30]

It is tempting to argue that the domestic sphere to which women are relegated has its own power, and to an extent this may be true. But it does seem to depend on the value given to women as autonomous selves and the way men use or abuse their physical power. Paul discusses the power of women through menstruation and child-birth in a Guatamalan village, yet there is still an overwhelming dominance of men in general and sexual brutality and aggressiveness, which brings forth the comment from one old woman that 'all men are like dogs'.[31] Parker and Parker also point out that 'principles underlying social structure are usually male centred', that women do not have power over the resources of societies and that they are sexually controlled.[32] They cite a cross-cultural study of domestic power in forty-one societies which found that in 66 per cent of cases wives deferred to their husbands, while in 7 per cent wives were dominant. In the remaining 27 per cent spouses appeared to have some equality.

While Tiger argues that experiences in the kibbutz in Israel show sex role stereotypes emerging, even after attemps at egalitarian roles,[33] Hacker re-examines the data and the assumption that women's biological drives made them 'choose' feminine roles. She found that the original equality was superficial and the kibbutz failed to produce equality because it identified 'equality with the masculinization of both sexes'.[34] Hacker then looks at other societies' attempts to pose as egalitarian, for example in the Soviet Union and China, and finds that the emphasis always lies with changing women's role but not men's. Thus, women are encouraged to work *more* but men are not forced to pick up an equal amount of work within the domestic sphere. The point of these anthropological and cross-cultural studies is that difference abounds, with as much valid evidence *against* the 'naturalness' of woman's powerlessness as can be gathered *for* it.

Men have attempted to define woman's real affinity as with the natural rather than the cultural world. This is by virtue of her biology and her procreative capacity. The argument is that woman's body and its functions are more involved with nature as it is 'more involved more of the time with "species life", while man is distanced from it physiologically'.[35] He thus has the time and freedom to create culture. Men have also tied woman's body and its functions deterministically to her social roles which are deemed to be of a lower order compared to man's cultural roles: she is relegated to the domestic sphere and therefore does not have power to create or change institutions of power. Finally, these social roles and her biology create a psychic structure for women which is supposedly closer to nature. Thus women are concerned with personalism and particulars, and with the concrete, while men work with ideas and the abstract.

Thus women are said to exist part-way between nature and culture, but are rooted in the former. Some exceptions occur in different periods when woman becomes closely aligned with culture. Ortner cites Nazi Germany where women—but only those of the 'chosen' kind—were to be the guardians of culture, and the Sirionö of Brazil who see male as nature and rawness, and female as culture. It seems that this distinction is brought up, however, only when women are called upon to *control* men—in matters sexual, for example. The picture of the rapist as beast, and woman as responsible for exercising control over him, allows man to blame woman for failing to control himself. Man seems only to equal nature when it allows him to excuse brutality in his behaviour. Moral (cultural) woman should then have controlled him.

It is generally true that men are more aggressive than women, and women more adept with personal relationships. But these differences spring from the influence of culture, ideology, socialization and the institutions of patriarchy which have defined woman's self through her biology. Child-bearing then becomes a dependency relationship for women as well as children, when different structures could make the experience one of maturity and life-expansion. The attempts to root woman's self in biology and nature is an attempt to justify the use of male power in controlling women.

Woman's body, man's myth

Male attempts to control and define women through the definition of woman's body and body-image are exemplified in its representation as fearful, problematic, dangerous and dirty. The myth of woman's body has worked to distance women from their bodily selves, leading them to accept this negative definition and allowing, for example, male-dominated medicine to mutilate them in the name of repair and healing.

Images of woman have been dichotomous. Contrasts between the light and dark lady, the madonna and the whore, the Victorian woman and the prostitute, are well known. These representations revolve around woman as fertile, as mother, and as sexual being. It has been asserted on the one hand that woman is appointed to observe moral rectitude, as represented by the Virgin Mother: yet on the other hand she has been denounced because of her wantonness and insatiable sexuality, which need punishment and control. Between these two extremes have lived the real women who struggled to survive the imposition of these images.

Myths are ways of knowing and are created so that man can control his world. Women are constructed in myth as the source of all evil: Eve seduced Adam, resulting in the fall of mankind, and Pandora released all the evils of the world. Woman is, however, a necessary evil, needed to 'perform the functions of sex object and child bearer', but otherwise unimportant and 'rightfully excluded from the company and affairs of men'.[36] The representation of woman as evil allows man to justify punishment and abuse. A good example is the witch burnings, when women were tortured and burnt, accused of insatiable lust and of cojoining with the devil.

The seductress is arousing but threatening. This threat is represented in fear of the vagina and in the myth of the castrating female.

Behind the fear of the vagina as a castrating agent lies the threat that the sinister female genitalia will control or devour. For example, the vagina is not uncommonly referred to as a vice with an unyielding grip or a gaping hole which draws, sucks, and devours the man.[37]

In North American Indian cultures, for example, there are at least twenty-two versions of the myth of the young girl with thorns around her vagina which the young man must break before marrying her. Many drawings symbolically represent the vagina as thorned or with teeth.[38] Man's fear of castration is clear in this representation. And it does not belong solely to the past. In his poem *The Book of the Body* Frank Bidart writes:

'I don't blame my wife—I was gone
a long time, and like everybody
else in Vietnam I did a lot of fucking around'.
He looked frightened and embarrassed, seeming to want me to
reassure him . . .
            I asked him to tell me about Vietnam.
'Anything you touched might explode. I know guys
just kicked a rock, and got killed . . .
            Once a buddy of mine
was passing a hut, when a gook motioned to him to come inside.
Inside a woman was lying on her back, with
a pile of cigarettes next to her. He threw
some cigarettes on the pile, got on top of her,
and shoved in his prick.
            He screamed.

She had a razor blade inside.
The whole end of his thing was sliced in two . . .[39]

Woman is also seen as fertile and responsible for creating. She is related closely to nature and to the cycles of the moon. But this too is ambivalent as 'the mother who has the power to give life also has the power to take it away'.[40] Ishtar, the goddess of fertility, is also the goddess of bloodshed and destruction. Fear of the destructive mother can be seen in the play *Medea*, where Medea kills her children after her husband's infidelity, punishing him by withdrawing life from his children and therefore immortality from him. In Australia, the Azaria Chamberlain case exemplifies this fear of the powerful, life-giving, death-giving mother. Here a mother was accused of killing the baby Azaria at Ayers Rock, which is a sacred Aboriginal site. Images of sacrifice and blood continually filled the newspapers, and Lindy Chamberlain was presented as 'unnaturally calm' and not emotional as a bereft mother should be. The media unconsciously expressed the fear of the mother—creative but dangerous and destructive.[41]

The powers of fertility were linked to the moon's cycle, which could also represent madness—the lunatic. When people did not know the origins of life, a woman's fertility was seen as mysterious and related to nature. Woman linked man to the earth. She was thus 'part of that nature which he could not control, which could destroy him with her capricious whims. To effect a separation of the mortal woman from the identity he feared she had with that power, he had to neutralise her magic'.[42] Male-defined science is a representation of the goal of control of both nature (the environment) and women's bodies (reproductive technology and gynaecology).

Because of her generative powers and her monthly bleeding, woman was and is seen as mysterious, enigmatic and unpredictable. Archetypes have been created and became part of the collective unconscious of man which Jung discusses. He describes it as 'the storehouse of latent memory traces inherited from man's ancestral past', and 'a residue that accumulates as a consequence of repeated experiences over many generations'.[43]

Because woman's bleeding was so mysterious—the wound that did not lead to death—taboos were developed around it which excluded women from contact with men either sexually, or through touching their food or eating with them. Adrienne Rich discusses the question of whether these taboos were initiated by women or men. Women may have created them out of respect for their own mysteries and in order to keep male sexual demands in control at least during one time of the month. But other writers argue that men's fear of women's fecundity and menstrual blood led them to create control mechanisms to protect themselves from her supposed powers.

The labels of 'unclean' and 'polluting' attached to menstruation also indicate a male-biased definition, as women would be unlikely to so name their own mysteries. Treatment for menstruating women has ranged from banishment to menstrual huts, to purification (like stepping over the fires of burning reindeer skin in Siberia), to direct punishment. If a woman failed to warn the man of her 'uncleanness' she could be killed, as she could be if she touched male possessions in the Iroquois and some Aboriginal tribes.[44] At the turn of the century women in Greek Orthodox religions were forbidden to take communion when menstruating.

Many of these social taboos still exist. Modern taboos make most people unlikely to have intercourse during menstruation. Women of pre-World War II generation were warned not to wash their hair or they would develop migraines. Many still call menstruation the 'curse'. An interesting recent public occurrence which showed the taboo to be alive and sick was the reaction to 'toxic shock syndrome' which was linked to the use of tampons, although some men also contracted it. Women were remonstrated with and lectured about 'keeping clean', and washing hands and bodies regularly. No word was mentioned of the fact that tampons are treated under cosmetic regulations, and are not produced and packaged in sterile conditions. The assumption was that toxic shock was related to menstruation, women's bodies, and therefore uncleanness. Women have been inculcated with attitudes towards menstruation which encourage them to look on it as unclean and a disability, something to be hidden.

Many of the symbolic representations of woman present her as the castrated man with a bleeding wound, and this is one of the origins of men's anxieties. They project their fear of castration onto women. After all, their genitals are external and therefore more vulnerable. By seeing women as castrated, men can reassert their superiority (their genitals are still intact), and it 'calms his castration anxiety by making woman, not man, the victim'.[45]

The fear of women which men have could originate in their envy of woman's body and fertility. This is evidenced in the ceremonial splitting of the underside of the penis in some New Guinea, Australian, Philippine and African tribes, in order that the splayed penis should resemble the bleeding vagina.[46] Psychoanalytic theories of womb envy such as those of Karen Horney and Bruno Bettelheim fit with these actions.

Wolfgang Lederer suggests that male 'counter-terror' through 'male counter-magic' (which women would call mutilation, abuse and misogyny), has been readily used in many lands in many ages to combat man's envy of women's perceived power. He writes that wise men are silent but that some 'admit to envying girls their breasts, their genitalia, their ability to bear children. And they hate them for it and dream of violence: cut off breasts, tear out vaginas—one wonders how much sadistic crime against women derives from such envy.'[47] Lederer also comments that man feels 'variously frightened, revolted, dominated, bewildered and even, at times, superfluous'. He fears being superfluous, obsolete, and representative of destructive forces, while woman lays claim to the life force: 'we diminish her power, to reduce her importance ... whereas in truth ... we need her, and depend on her altogether'.[48]

Leader thus discusses the 'why' of male fear, but Ehrenreich and English, and Mary Daly, discuss the processes—the 'how' of man's 'counter-terror'. Daly stresses the sexual brutality of man, concluding with the misogyny of gynaecology.[49] Ehrenreich and English trace the medical profession's attitudes to the treatment of women, showing its increasing power over women's bodies.[50] The structures within patriarchy which exercise control over women's bodies continue to feed on those feelings of fear, envy and hatred from man. Bruno Bettelheim has shown that 'males have everywhere tried to initiate, annex and magically share in the physical powers of the female'.[51] Modern gynaecology and obstetrics have proceeded to take women's bodies and the process of birth from them and to make it man's. Within the new field of reproductive technology, womb envy and the desire to control are expressed in the increasing male control of pregnancy and of conception itself.[52]

The effect which this has on women is that they devalue themselves. Their image of their physical self becomes a negative one, and they are embarrassed and ashamed of its processes, which are seen as sickness. This negative self-image contributes to a perpetuation of that picture of woman as unclean and dirty, and purely as a (malfunctioning) biological self. She continues, for example, to cover or expunge body odours, and is shamed into using vaginal deodorants. Menstrual taboos have helped to create and perpetuate social and psychological structures designed to devalue and negatively define woman's body. In our society, any attempt to break out of its chains are met with male disgust and fear.

While rendering the experience of menstruation as dangerous and disgusting, men have not failed to problematize women's biology in other parts of the cycle. Pointing to the existence of some pre-menstrual behaviour and mood changes in women, they have generalized that women are unpredictable, unreliable and dangerous when they are *not* bleeding. Pre-menstrual 'tension' (PMT) or 'syndrome' (PMS) is not new, nor are the arguments about it, nor the social context in which it is argued. It does have serious political implications for women. The term itself was coined in 1931 to describe a set of personality changes supposedly associated with the pre-menstrual drop in oestrogen and progesterone levels in a woman. It describes a period of high irritability, depression, anxiety and feelings of low self-esteem. Violent crimes by women and suicide have also been found to occur more during the pre-menstrual and menstrual days. However, the general level of crime and suicide for women remains markedly below that for men and this is rarely stressed.[53] Some of the research findings to date are as follows: among women who commit or attempt suicide, 22 to 36 per cent are estimated to do so during menstruation: 53 per cent of women's attempted suicides occur during the four pre-menstrual or the four menstrual days, as do 49 per cent of acute medical and surgical admissions and 52 per cent of accident admissions to hospitals.[54] A study by Moos found that approximately 30 to 50 per cent of responses from 839 'normal young married women' to a questionnaire indicated cyclic symptoms in irritability, mood swings, tension and depression.[55] Notably, none of the researchers discuss the experiences of the remainder of their unaffected respondents who are usually in the majority.

Problems abound with the methodology of research on PMT, and the following are some examples. The women in the samples studied are often those who are easily accessible because they are seeking medical attention for a complaint, so it is not surprising that they report more distress related to menstruation than do healthy women. Women who answer advertisements for questionnaire respondents are also likely to be those who have problems which they want to discuss. If a woman is then asked to complete a 'menstrual distress question-naire', she gets a clear picture of the information being sought and responses are already biased. Many studies use retrospective ques-tionnaires in which the respondent has to recall information, which may make the information unreliable.

The symptoms studied are mostly couched in negative terms, and therefore the results will be skewed. And, the *degree* to which a symptom is experienced is impossible to assess. Mild irritation for one woman might be experienced strongly by another. Often the only parts of the cycle studied are the pre-menstrual and menstrual phases, so there is no base line with which to compare experiences. Furthermore, Margi Ripper points out that the research is based on the assumption that all women have 28-day cycles, which is not accurate, so locating the phase of the cycle is itself difficult.[56] There are few comparative studies which compare women in the pre-menstrual phase with those who are not, and there are no studies comparing them with men.

In a detailed study Paige has discussed the influence of social factors on the experience of PMT and menstruation and found that the hormonal influence did not account for the whole of a woman's experience of menstruation. Religious beliefs were influential and women with traditional attitudes to femininity were more likely to be affected than working women.[57] Estelle Ramey has also noted with respect to menopause that 'those with continuous ego-satisfying work, suffer menopausal symptoms much less'.[58] Chafetz looks to social experiences as influential in women's attitudes to PMT. She writes that it is possible 'that females have been told since childhood that they can expect to experience these different mood swings during their cycle, during menopause, and after childbirth . . . If this is the case, cultures, not hormones create the emotional variation'.[59] Con-sidering the taboos relating to menstruation it is clear, as Paige suggests, that 'it is no mere coincidence that women suffer anxiety and distress during a biological event they consider embarrassing, unclean—even a curse'.[60]

The politics of PMT becomes a two-fanged dragon for women. Many women say that they do *not* experience PMT and their experiences must be accepted as valid. Some women say they *do* experience changes. If it is accepted that some women do experience PMT, men use that as a justification of their representation of women as unstable, neurotic and in need of control. (The assumption here is that stability is the norm.) They can exclude women from some spheres of activity, for example within the job market, because they are unreliable due to their biology. The debates over the admission of women airline pilots in Australia involved this justification for exclusion.

On the other hand, if it is argued that women do *not* experience PMT, the experiences of those who say they *do* suffer intense discomfort and pain are invalidated. This reinforces the picture of neurotic, bored women whose problems are 'all in the mind'.

But women should monitor closely this burgeoning interest in PMT. One American doctor has already expressed concern at the 'legitimizing of PMT by some sections of the medical community'. She discusses the evolution of menopause as a so-called health problem. 'By the 1970s, a normal life cycle progression had turned into a supposed syndrome, which finally became a disease called oestrogen deficiency'.[61] The history of the relationship between medicine and women makes it wise to be wary of these attempts to name and create women's biological selves and experience of those selves. Often, women who claim to be symptom-free are ignored or said to be unable to detect their own pattern of behaviour. One group of scientists claimed:

It is estimated that from 25−100 per cent of women suffer from some form of pre-menstrual or menstrual emotional disturbance . . . Eichner makes the *discerning* point that the *few* women who *do not admit* to pre-menstrual tension are *basically unaware of it*, but one needs only to talk to their *husbands*, or *co-workers* to confirm its existence.[62] (emphasis added)

In a thorough analysis of PMT, Laws et al. clearly show how it is being used to discredit women, threaten their independence and exert social control.[63] Laws traces the history of the concept and likens it to the use of clitoridectomy in the nineteenth century for very similar 'problems'. Laws and Eagan expose the dangers of medical therapy using progesterone, a remedy with known side-effects and no proven record of success for a syndrome yet to be shown to exist, but one on which large and successful drug companies are based.

They stress that women's subjective experience is not in question. But the way those experiences are taken by the patriarchal medical profession, named as illness and 'not normal', then medicalized, is questionable. As Laws writes: 'It is clear that some women feel those changes more intensely than others, but those changes still constitute part of a woman's being, and are not signs of sickness.'[64] PMT, she argues, can be symptomatic of other problems in women's lives. For example, it may justify their demand for someone to listen to them, women being made constantly invisible; and stops women being labelled as neurotic if they are angry. Any emotional changes can be called PMT problems and women can be 'helped' by medicine to become the passive, well-behaved women they should be. But why *are* they distressed, unhappy or angry?

The scientific evidence for PMT is lacking. Editorials in the *British Medical Journal* and the *Lancet* state that it is impossible to distinguish PMT from the experiences of a normal woman. One of the strongest proponents of pre-menstrual tension in England has been Dr Katharina Dalton, who makes a successful living from using hormone replacement therapy for PMT and menopause. She has also become well known through her court appearances in support of women on criminal charges who claim PMT as their motivation.[65] Yet Dalton herself has trouble identifying the 'true sufferer'.

Women suffer from over-monitoring of changes in their bodies because they are constantly taught how inadequate and faulty their bodies are. But whose interests does PMT serve? Those of medicine and drug companies certainly; and men whose wives are 'problems' for them, and a few women for whom 'therapy' works. But women in general gain little. They are fed dangerous drugs, encouraged to see strong emotion as a 'problem', become over-anxious about changes which are normal to most women, and are encouraged to seek medical help to *control* their bodies and their selves.

Hey considers the two cases of women found not guilty of violent acts, one murder, on the grounds of diminished responsibility due to PMT. Though it is clear that the social context was part of the problem (a fight with an abusive and drunken lover whom she then ran over), Christine English was acquitted on these grounds, creating legal precedence for the 'problematizing of women's bodies, and a non-problematizing of male behaviour'.[66]

Finally men's cycles should be considered in comparison with those of women. Are they the supremely stable bunch they portray themselves to be? Normality here is male-defined as equivalent to having no cyclic mood variations. But how 'normal' are men?

In 1931 Hersey studied male cycles, in the belief that male factory workers were incorrectly thought to be stable and unchanging in their daily capabilities. For a year he studied workers and management, concentrating on a group of men who seemed particularly well adjusted and at ease in their jobs. Using interviews four times a day, regular physical examinations and interviews with workers' families, he established charts for each individual. These showed that emotions varied predictably within the rhythm of twenty-four hours *and* within the larger rhythm of a near-monthly cycle of four to six weeks.

Low periods were characterized by apathy, indifference, or a tendency to magnify minor problems out of all proportion. High periods were often marked by a feeling of well-being, energy, a lower body weight and decreased need for sleep. Hersey asked, what should be done about this periodic cycle in men? He said they might attempt to eliminate the cycle, but this would mean being sure of its origins and he was dubious about the tendency he saw in society to insist on men 'without variations'. He suggested in contrast that they should accept 'the existence of periodic changes in mood and ability as something akin to a fundamental *law of our nature*' (my stress), and 'adjust our [men's] lives to its demands',[67] an interesting message for men.

The endocrinologist, Estelle Ramey, also produced evidence that men do have monthly cycles. 'The evidence of them may be less dramatic, but the monthly changes are no less real'.[68] These cycles are both physical and behavioural. A study carried out in Denmark over sixteen years looked at the fluctuating levels of male hormones contained in male urine. The result was evidence of a thirty-day rhythm in the ebb and flow of the hormones. Ramey also quotes evidence for the need to consider male cycles within the job situation. She writes of success in Japan on this front:

The directors of the Omi Railway Company of Japan, for instance, are pragmatic students of human behaviour and have therefore decided to accept the fact that men have lunar cycles of mood and efficiency. This company operates a private transport system of more than 700 buses and taxis in dense traffic areas of Kyoto and Osaka. Because their operations are plagued with high losses due to accidents, the Omi efficiency experts began in 1969 to make studies of each man and his lunar cycles and to adjust routes and schedules to coincide with the appropriate time of the month for each worker. They report a one-third drop in Omi's accident rate in the past two years, despite the fact that during the same period traffic increased. The benefit to the company has been substantial.[69]

In Australia in 1976 Margaret Henderson conducted research into men's cycles. She found a monthly temperature cycle with characteristics similar to that experienced by women in their menstrual cycle. The length of the cycle varied from seventeen to thirty-five days. She also found that men and women who live together, 'cycle' together. At the time of the mid-cycle temperature drop in the woman, there was a similar temperature dip in the man.[70] This is not unusual when we consider that many women who live together find their cycles gradually coinciding so that they menstruate at similar times. McClintock has found this in women who live together, but also in women who work together and the cause is still not clear.[71]

The evidence of biological rhythms affecting men's temperature, hormones and behaviour is there; but little is heard of this and the subject is not pursued with vigour in the research world. It is threatening to the whole concept of male stability in personality and behaviour, which ensures man's superior position to woman and which can be used to exclude her. Men have concealed their own variable nature while exaggerating women's, tying her again to the biological, to the immutable. Ramey comments on the possible response if this vision were successfully questioned:

Such a departure from the mythology of male biological stability might produce in men the same kind of psychological wrench that Copernicus inflicted on them when his theories revealed that man is not, in fact, the center of the universe ... Women's chains have been forged by men, not by anatomy.[72]

Broader implications of the chains of anatomy

The tying of woman to her biology in a way in which men are not has had extensive political implications for women, causing, for example, their exclusion from certain jobs in the paid workforce. Historically, it was used to deny women an education. In England, for example, Darwin stated categorically that 'man is more powerful in body and mind than woman'.[73] Man's superiority came from his struggle with 'life'. With the power of modern evolutionary theory behind him, Maudsley argued in 1874 that the female sex was weaker, due to women's need for energy to reproduce. Higher education would damage these potentialities. Women needed to understand that it is

not a question of two bodies and minds that are in good physical condition, but of one body and mind capable of sustained and regular hard labour, and of another body and mind which for one quarter of each month during the best years of life is more or less sick and unfit for hard work.[74]

Science also proceeded to belittle the brain size of women and to show them how inferior they were. Geddes, in his *The Evolution of Sex* in 1889, expounded the need to 'follow the signs of biology'. Women would cease to be altruistic if they were not passive and if they sought 'masculine activity'. Conway writes:

Lest women should rail against a biological providence which had given them 'habits of the body' which could be exploited by stronger men, Geddes hastened to proclaim that the study of sexuality in human societies gave no hint that political or social factors had led to the subjection of women.

Women should not attempt to alter their biological destiny as mothers by interfering in the world of men. Their personalities and true nature, dictated by their bodies, made this unnatural. 'This typology of biologically-determined sexual temperaments expressed unaltered the romantic myth of the rational male and the emotive, intuitive female.'[75]

In America too, the debate raged, as the medical profession insisted that intellectual activity would drain the reproductive organs of their power, and lead girls to produce weaklings and deformed children. A woman was therefore the prisoner of her reproductive system, though of course 'the male reproductive system ... exerted no parallel degree of control over man's body'.[76]

When reading accounts of the prattlings of the medical profession and of the biologists in this area, one sees clearly the intention to deny women education and confine them in stupidity to their domestic role. Classist assumptions were also inherent in these debates. Feminists continually pointed out the extremely hard labour which working-class women did during menstruation, only to be directed to examine their children as evidence for the medical profession's argument. It is a mark of women's resistance that they managed to open schools and to enter unversities, even if at that period the degrees were not always granted them at graduation.[77]

By stressing women's body and biology as 'abnormal' and constantly in need of repair and rearrangement, men have structured a system where experimentation on and mutilation and abuse of these bodies can be justified. The linking of women's biology to their personality has also justified the treatment of women with psychological problems through biological change. In nineteenth-century England and America this was particularly true. One of the major medical practitioners of the day was William Acton, who stated that the 'majority of women are not very much troubled with sexual feelings of any kind'.[78] Some women, however, were found to be plagued by a variety of indispositions which needed a surgical cure. Masturbation, nymphomania, and rebelliousness of character could be cured by clitoridectomy, and a range of disorders—troublesomeness, overeating, erotic tendencies, persecution mania—could be cured by ovariotomy (or female castration).[79]

Clitoridectomy was the first operation created in Western countries to control women's mental disorders.[80] In England the inventor and main proponent of clitoridectomy was Dr Issac Baker Brown who wrote a book outlining his technique, which involved excising the clitoris with scissors under chloroform. The book caused an outrage and he was expelled from the Obstetrical Society because, in the words of the President of the Medical Society of London, 'It is the manner in which the operation is performed, *not the operation itself*' (emphasis added).[81] The operation fell into disrepute in England; however it continued to be popular in the United States between 1870 and 1904, perhaps even until 1925. Ovariotomy was particularly popular because of the need to control the sexual temptress who would weaken a man's masculinity. It too was used to 'treat' 'mental disorders' in women.

Mainly middle- and upper-class women were victims of these medical atrocities; their husbands could afford to pay for the 'treatment'. However, it was the poor and black women who were the guinea pigs for medical experimentation. In North America, pioneering gynaecological surgery was conducted in the 1890s by Dr Marion Sims 'on black female slaves he kept for the sole purpose of surgical experimentation'.[82] On one of these women he performed thirty operations in four years. After he moved to New York, Sims used immigrant Irish women in the wards of the New York Women's Hospital as experimental subjects.

This situation was true also for poor women in other areas of medicine. Adrienne Rich discusses the history of motherhood and the ways in which women lost control of midwifery after about 1663. As the control of child-birth passed into male hands, death from infection increased and the methods of delivery took on a form of brutality. Forceps were called the 'hands of iron' and the victim of one of the major test cases was a dwarf woman, deformed and incapable of delivering a child possibly conceived through rape, who died after more than three hours had been spent 'working her over' with forceps. Rich comments that she 'died in torture'.[83]

Experimentation on women's bodies continues today, as drug companies test drugs on live subjects. Women on reproductive technology programmes are being used as living laboratories for new techniques designed to give masculine science control over procreation and conception. In the process, they are being given dangerous and untested fertility drugs, which may have caused cancer in a number of cases documented so far.[84]

Examples of the mutilation of women still abound. Gena Corea discusses American Dr James Burt, who carried out what he calls 'love surgery'. He believes that the female anatomical system is faulty because women do not orgasm during penile penetration. Therefore he reconstructs the woman's vagina and genitalia, dragging the clitoris closer to the vagina. When Burt wrote his book in 1975, 4000 women had so far been treated by this man, many of them without informed consent. Burt himself has written that 'in many hundreds of these patients, the patient had not been informed that anything more had been done to her than delivery and episiotomy and repair' after the birth of her child.[85] So Burt has 'redesigned' the bodies of hundreds of women without their knowledge and continues to do so. Even though he is shunned by his colleagues, the medical profession has done nothing to stop him from performing this surgery. It would seem that there is an inherent contradiction between the operation of medicine and the concept of women's control.

Woman's body-image is shaped by these experiences and attitudes. It is 'problematic', in need of healing or repair. But her body is also presented to her as having a purpose *for* man, not for woman herself. Women's bodies are used to titillate and reinforce male sexuality continually, in film, in advertising, and in pornography, and are variously admired and abused. Body shape itself is created in an ideal form for women to emulate. In Australia, the Big M advertisements were used by the Victorian Milk Board to promote milk, while also promoting the image of the thin-bodied, large-breasted nymphs which women should aspire to become. In reality few of these exist, so failure to live up to the ideal is built into the image. And this objectification of women's bodies leads to the objectification of women themselves: they become again the 'other', the 'non-person'.

The image women have of their bodies even affects their eating behaviour. Discussing her experiences as a woman endeavouring to fit the stereotype, Kim Chernin writes of the abnormal attitudes to eating that women develop in order to achieve the appropriate shape. She discusses her own re-learning of the pleasures of normal eating:

My body, my hunger and the food I give to myself, which have seemed like enemies to me, now have begun to look like friends. And this, it strikes me, is the way it should be; a natural relationship to oneself and the food that nourishes one. Yet this natural way of being does not come easily to many women in our culture.[86]

One interesting element in the current pressure on women to fit the stereotype is the stress on *smaller* women and *larger* men as the women's movement has increased its impact. Chernin argues that the pressure to be childlike and small has a deliberate political purpose. Reissman also makes this point, comparing the height to weight tables for the sexes from the US Metropolitan Life Insurance tables in 1959 and 1979. In them, the desirable weights for women have decreased, but for men they have generally increased.[87]

There is a double message women get about their bodies. On the one hand, they are supposed to cover them in order not to incite men. For example, young girls must have their breasts covered. On the other hand, once breasts are developed, they are not meant to be totally concealed, but should be 'tantalizingly outlined under clothing as a sort of promissory note to be enjoyed by men'.[88] Women are frequently worried about their breasts because they are 'often categorised and labelled on the basis of their cup sizes'.[89]

Breasts, and women's attitudes to them, are doubly important when we consider the medical profession's overwillingness to carry out a complete mastectomy as soon as a lump appears in a woman's breast. Though this is sometimes presented as essential treatment, there is doubt within the medical profession itself as to the necessity for such radical surgery.[90] A British woman, and ex-nurse, Jill Louw was shocked at the 'emotional blackmail' used by a doctor she saw to force her to accept a mastectomy. Having the privilege of contacts in the field, she found out that the professional literature was of 'the overwhelming opinion that a mastectomy was not necessarily the answer'. Pressure was brought to bear on Jill Louw and her husband, but she wanted and obtained a referral to a woman consultant surgeon who discussed a variety of possible treatments. Finally she opted for radiotherapy and although the first doctor had given her only two years to live, when she was interviewed in Australia in 1983 she had lived three without the radical surgery.

Louw's report from the first surgeon described her as 'difficult and emotional' and her husband was told 'women become very emotional over these things'. This was the approach taken by male surgeons to a woman who simply wanted to wait, think and discuss the issue. Her husband commented on the undue haste evinced by the doctors and their attitude to him which was 'to enlist me as a male ally in trying to persuade Jill to have a mastectomy, as though I had the right over her body'.[91]

The publishing of this story produced a feverish response from medicos and Kathy Kizilos compiled a report soon after for the *Age*.[92] This indicated that most doctors 'opt for mastectomy', even though studies indicate that a variety of factors can influence the need or otherwise for surgery, such as whether the cancer has occurred elsewhere in the body. For many women the operation is trauma-free and their one concern is to eliminate the cancer. But for some, it may be experienced as mutilation. An interesting comment was made by Lin Layram in *Half the Sky*: 'I still feel uneasily that somehow it is a punishment for my sexuality and pride in my body in that period before the operation'.[93]

Women's body and biology, negatively defined, create a negative self-image and a lack of faith in the body as strong and healthy. As Jeanne Gullahorn comments: 'in our culture growing up female can easily mean developing an incomplete body image or one that focuses on the "necessity" of making sexual elements acceptable to others'.[94] This stress on unattainable ideal physical standards leads women to self-doubt and self-dislike.

Breaking the chains

Women need to revalue their physical selves. Feminism has encouraged this revaluation through encouraging women to gain knowledge of their bodies and therefore control over them. Women have resisted the chaining of their selves to a biological base with its implicit assumption that the power differentials between women and men are justifiable by reference to 'nature'. This is done while not dismissing the importance of the body to a person's sense and experience of self.

Medicine is male-owned, operating to control women, often to the detriment of their health. In the late 1960s the women's health movement gathered momentum, developing since then an international scope with diverse approaches to women's health. Services run by women, for women, have revised the way women's health has been viewed, stressing prevention rather than a reliance on hi-tech, expensive and dangerous technologies and drugs.

Lisa Tuttle has indicated that the work done in this movement has covered three areas: the education of women through changing conciousness and the availability of information; providing health services to women; and struggling to change established health institutions.

One of the landmarks of the women's health movement was self-help gynaecology. In April 1971, in Los Angeles, Carol Downer showed women for the first time how to use a speculum to examine their own vagina and cervix and the bodies of other women. In these actions, women came to see for the first time inside themselves. Women's bodies were no longer solely for the male medical gaze. These actions demystified women's bodies themselves, and made the gynaecological ritual more obvious in its humiliation of women. As Ellen Frankfurt wrote:

I hate to use the word 'revolutionary', but no other word seems accurate to describe the effects of the first part of the evening. It was a little like having a blind person see for the first time—for what woman is not blind to her own insides? The simplicity with which Carol examined herself brought forth in a flash the whole gynaecological ritual; the receptionist, the magazines, the waiting room, and then the examination itself—being told to undress, lying on your back with your feet in stirrups ... no-one thinking that 'meeting' a doctor for the first time in this position is slightly odd.[95]

Women's health centres continue to provide a threat to patriarchal medicine and many of them are raided, women being charged with a variety of trumped-up charges. For example, the Feminist Women's Health Centre in Los Angeles was put under police surveillance in 1972 and in that year Carol Downer was arrested and charged 'with practising medicine without a licence, for the specific crime of helping another woman apply yoghurt to her cervix as a cure for monilia'.[96]

In 1969, when little information was available on women's health, the Boston Women's Health Book Collective put out the first edition of *Our Bodies, Ourselves*, which became a basic reference text for women all over the world. The second edition, published in 1985, has continued this tradition with an expanded view of women's health and the medical system which attempts to control it. Stressing preventive measures, and the need for women to understand the workings of their bodies, this book is an act of resistance against misogynist health 'care' throughout the world.[97]

Women's biology and bodies are integral in their definition of self, negatively envisioned by men. Walum has synthesized the meaning of that bodily self when she writes: 'Our sense of our bodies—our knowledge about them, the pleasure we derive from them, and the joy or distress we experience when we contemplate them—is the basis upon which we build our self-image'.[98]

2
The Rational/Intellectual Self as Expressed in Science

Scientific knowledge and the construction of woman

The dichotomy between nature and culture has served men well. If man equals culture and woman is nature, she has no need of the intellect but is primarily responsive to her body's needs. Man has presented himself as the creator of ideas, and woman's sense of herself as intellectual and rational has been eroded. She has become the irrational, emotional, subjective sex, full of carnal knowledge. Man, on the other hand, has made his image one of the rational, logical and thinking sex, elevating these characteristics as distinguishing him from the animals. Woman he defines as closer to the animals because of her inescapable bodily functions of menstruation and pregnancy. Man seeks to control 'irrational' nature by a variety of means, one of which is the construction of reason as superior.

In terms of selfhood, reason has been represented by men as a transcendence of femininity. Reason has been named as expressing 'the real nature of the mind', and has been incorporated into 'our criteria of truth' but also, as Genevieve Lloyd writes, into 'our understanding of what it is to be a person at all, of the requirements that must be met to be a good person . . . .'. She goes on to say:

The obstacles to female cultivation of Reason spring to a large extent from the fact that our ideals of Reason have historically incorporated an exclusion of the feminine, and that femininity itself has been partly constituted through such processes of exclusion.[1]

Man has expressed his determination to reify reason in his creation of science. It is rational, objective and concerned with Truth. It has been 'defined as *the* expression of the male mind: dispassionate, objective, impersonal, transcendent. The female mind—untamed, emotional, subjective, personal—is incompatible with science'.[2] Scientists have treated the natural world, and woman as part of nature, as objects of study. Their findings have been used to control and define woman, justifying her oppression by labelling her as unstable and inferior.

While so constructing scientific knowledge, the scientific establishment excluded in its practice the possibility of women's input into this knowledge construction. So it has fed ideologies of oppression while its own structures maintain the exclusion of women. Writing of the erasure of women as intellectuals, Dale Spender notes that men have

claimed for themselves a monopoly of the mind, they have described and legitimated themselves as the 'authoritative' sex, the sex with the capacity to be 'objective'... Women who reveal their intellectual resources are often described as having 'masculine minds', which is a clever device for acknowledging their contribution, while at the same time it allows it to be dismissed, for a woman with a 'masculine mind' is unrepresentative of her sex, and the realm of the intellectual is still retained by men.[3]

Whenever women question their relegation to the non-rational, non-thinking class, men can always point to science and ask, 'If women are so intellectual, why are there no great women scientists?' (always with the exception of Marie Curie!). The answer given to women has two elements: women cannot achieve within science because they are constitutionally (by nature) unsuited to it, and because they do not *want* to do so.

If women believe themselves to be less intellectual, rational or thoughtful than men, two primary effects follow. First, in terms of selfhood and identity, women have a less positive sense of self, a weaker faith in their own opinions and a general lack of confidence in themselves and in women as a social group. Second, in practical power terms, women have less access to an important power base in society. For example, in science, they have less power in defining and constructing knowledge, and in a practical sense, reduced access to engineering and technical skills and (as technology increases in the workplace) less access to jobs.

Historically, the construction of scientific knowledge has contributed to the definition of woman. Far from being 'objective' it has served men in their oppression of women. As Galileo and then Darwin's explanations of the world became accepted, the usefulness of religion as authority waned. Science became and has remained the new authority, the new 'truth'. The emerging world view of the new scientific age was 'masculinist', seeing woman as alien and inexplicable. The 'Woman Question' became paramount, and science objectified 'woman' as an area of study. Ehrenreich and English write:

Everything that seems uniquely female becomes a challenge to the rational scientific intellect. Woman's body, with its autonomous rhythms and generative possibilities, appears to the masculinist vision as a 'frontier', another part of the natural world to be explored and mined. A new science—gynaecology— arose in the nineteenth century to study this strange territory and concluded that the female body, is not only primitive, but deeply pathological.[4]

In her analysis of selected articles in the *Popular Science Monthly* between 1870 and 1915, Louise Newman has recreated the debate about the nature of 'woman', the dangers of educating or employing women, and the fitness of women to vote or become involved in politics. Science had become the tool of patriarchy in its control of women's bodies and lives.[5]

Science has been based on an ethic of domination and control, primarily the domination of nature, which has been represented as female. Francis Bacon called on men to turn their 'united forces against the Nature of things, to storm and occupy her castles and strongholds'. He urged them 'to bind [Nature] to your service and make her your slave'.[6]

Brian Easlea has shown that post-sixteenth century science has been primarily *irrational*. Though some benefits have accrued to people in general, science has mainly been used as a tool of suppression and violence, and his book traces the sadistic use of science against women. One of the irrationalities, Easlea argues, is for science to claim to be seeking to solve problems while it creates greater problems in seeking those solutions. For example, in gynaecology amniocentesis is now routinely used, yet there are dangers to both foetus and woman.[7] 'People are found—well, *men* are to be found—who explicitly state they wish always to live in a world where their heroic (scientific) problem-"solving" serves only to generate further and more serious problems for a united mankind to confront'.[8] This is a basically '"masculine" desire'. The scientific ethic, as Overfield has commented, like capitalism and imperialism, is 'based on exploitation, elimination of rivals, domination and oppression.'[9]

'Objectivity' has been the supposed hallmark of science, allowing it to claim a patent on 'truth'. But science is *not* the objective and value-free intellectual pursuit it poses as. It is tied intricately to society and reflects social (male-dominated) values. Male society determines what it is that is studied and in return, science reinforces social values.[10] Sexist research has developed sexist technology and together they operate to 'move control of women's lives from women to men of the dominant group.'[11] Bias intrudes in a number of areas. First, theories reflect fundamental ideological commitments which are rarely explicitly stated.[12] They reflect the values of the period in which the theory is developed. For example, in the 1950s when women were being forced out of the workforce and back into the home, science (in this case, psychologists) created 'maternal deprivation', since dismissed in a period when women are needed in employment once again.

Bias also intrudes through decisions on what is studied and what is not. So the negative aspects of the menstrual cycle are studied and re-emphasized repeatedly, while few study the creativity women experience during the cycle. Men's cycles are not studied, in case men too are found to be periodically 'unbalanced' and 'irrational'.

It is clear from the anthropological information discussed in Chapter 1 that the choice of sample population or group of people or animals for study can make a vast difference to the results obtained. A striking example for psychology was the development of the theory of achievement motivation. In studies used to develop this theory, women did not produce the expected results (that is, similar to the men), so they were dropped from the studies and from the theory. The reason they had not 'performed' was because of male bias in the tests used. The research was of 'high quality' and carried out in an 'objective' fashion. But over time many writers forgot 'that the results obtained in most studies applied only to men; they continued to cite the results as if they applied to everyone. But there was one problem: half the population had been systematically excluded from the research'.[13]

Within the 'harder' sciences, the opposite is often true. There, mainly all-women populations are used to test new drugs. In some cases this is done without their knowledge or consent. Side-effects are rarely discussed with women patients on the assumption that it would only worry them unnecessarily. The results of thalidomide, the side-effects of the pill, the deaths of women from infections due to the Dalkon Shield, and the cancer inherited by daughters of women who were given DES as a morning sickness pill, are a few examples of women being used as guinea pigs.[14] Recent developments within reproductive technology reinforce this perception of women as living laboratories.

Ruth Bleier, a professor of neurophysiology and women's studies, has written a critique of biological theories on women which she says amount to 'an elaborate mythology of women's biological inferiority as an explanation for their subordinate position in the cultures of Western civilizations'. She discusses science as a discourse, which like all discourses both reflects and generates relations of power. Discourses *produce* rather than *reveal* truth. Science, she argues, has not only constructed gender differences, dichotomies and dualisms, but has constructed *itself* to represent these dualisms. As a tool to maintain patriarchy, it has divided civilization into contradictory spheres, male and female, and has developed 'a dualistic mode of thought, the development of concepts and ideologies of oppositions, dominance and subordinance, culture and nature, and subject and object'.[15]

Science has developed a structure which is hierarchical and competitive. To progress, the scientist has to get grants for research and publish. This can only be done if the area to be studied is 'acceptable' to grant-givers, publishers and promotions committees. Ehrenreich and English have shown how money awarded by institutions like the Rockefeller Foundation moulded and created the medical profession early this century. Male bias in the way publishers and editors control publication of certain kinds of material and reject others has also been explored.[16]

The scientific mystique fosters the creation of experts to control our lives, masculine experts to replace the male god. A certain awe has been given to these men because their control of nature and technology makes them powerful. The veiled and unspoken threat exists that the knowledge they control may be withdrawn, leaving us with no protection, no 'progress'. Scientists also retain power through the creation of new problems to be solved, or as Easlea puts it, the new 'challenge'. It pays men to keep science in this position, conveniently placed on the culture side of society and necessary in the control of nature (woman).

Defining woman by her exclusion from science

The structures of science itself operate to exclude woman and by so doing, to define her as unstable and unfit for scientific work: a self devoid of rationality. Women are told they cannot achieve within science because they are constitutionally (by nature) unsuited to it, and because they do not *want* to do so. But what are the numbers in science and why are they so few? Do women want to be included? These questions lead to an unmasking of the way men have constructed women out of the scientific field.

How many women are in science?

Women are underrepresented in science. Statistics from the USA, Great Britain, Australia and New Zealand show similar results. Rossi analysed American women's representation in science from 1950 to 1960 and found the smallest percentage in engineering and physics and the largest (about 27 per cent) in biology and mathematics.[17] Zinberg found that by 1977 at university level, there had been an overall increase in female representation: women students represented 6.8 per cent in engineering, and in physics comprised about 27 per cent of students; mathematics and biology still had the highest representation—41 and 45 per cent.[18]

In Great Britain women predominantly study in the biological sciences and men in the physical sciences, engineering and technology.[19] Robertson also found this trend in Australia in school science enrolments.[20] Australian data indicate that the proportion of women students enrolled in science-based degrees is rising sharply, though more do medicine and veterinary science than do physics, chemistry, agriculture or engineering. At Sydney University in 1980, 39 per cent of science degree students were women and 33 per cent of medical students were women. Engineering, however, remains *very* low in female representation—6 per cent at under-graduate level and zero at post-graduate level. In science only 16 per cent of staff were women, and in medicine, 19 per cent.[21]

A frequent argument against allowing women into medicine is that they leave to have babies and do not return, therefore 'wasting' their expensive training. However, in a survey by Ione Fett on the whole population of Australian women medical graduates (2540), 89 per cent were employed compared with 92 per cent of medically qualified men.[22]

The following table gives some indication of the differences in representation in medicine and engineering in four countries.

Percentages of women among doctors and engineers in various countries[23]

Country	Doctors	Engineers
Britain	17	0.5
USA	8	1
USSR	74	30
Australia	16	0.2

Though female representation is high in the Soviet Union, these fields are low in status in the USSR, supporting the argument that predominantly female occupations are systematically given lower status.

Keir has noted that in New Zealand the government employs the majority of scientists engaged in research and development, and only 6 per cent of these are women. It is interestng too, that the latest survey (until 1979) of salaries by the New Zealand Association of Scientists found that of the 2553 scientists who completed the questionnaires, only 265 (about 11 per cent) were women. As with studies of a similar kind in America, the women scientists tended to have lower qualifications than men—fewer had doctorates. Most men were in research, while the women taught. Thus in the Keir study, 39 per cent of women scientists taught in secondary schools, compared with 11 per cent of men, while 51 per cent of men worked in government positions compared with 28.4 per cent of women.[24]

In Keir's study there was a difference in the marital status of women and men scientists. In the age range of 20−64 years, 86 per cent of the men were married while only 59 per cent of the women were. In the general population the figures would be 81 per cent and 88 per cent respectively, so while men are close to the norm, the women are not typical. Fett found similar results in Australian medicine. Considering the marital status of her 1632 women and 1414 men graduates of the years 1920−1969, she found that 22.3 per cent of women compared to 6.9 per cent of men had never been married; 69 per cent of women and 89 per cent of men had married once or more; and 3.9 per cent of women and 1.9 per cent of men were separated or divorced.

Keir also points out that a majority (54 per cent) of married women in her New Zealand group had no children, while less than a quarter (22.5 per cent) of men had none. Women in science feel acutely that it is impossible to advance within the field while rearing children.[25] Rossi indicates that the American situation is similar, with women scientists less likely than the men to be married and less likely to have children.[26] Many women scientists are choosing either not to have children or to delay child-rearing until their work is advanced. No structure has been established to enable women to do both. If a woman leaves the workforce she loses experience and knowledge in a rapidly changing field, and few retraining and re-entry courses are offered. Moreover, she will fall behind in the race to publish. Thus women scientists are even more disadvantaged by marriage and children than women in some other occupations. This often leads them to the more flexible teaching jobs.

As Fett's study of doctors suggests, child-rearing and domestic duties still fall to women, and 'all evidence suggests that . . . the woman . . . expects, and is expected by others, to take on this role.'[27] Recently, women doctors in the US were surveyed. Pregnancy was difficult for those women who were completing their residency (four-fifths of those involved in the study), as most residency programmes did not have maternity leave provisions. Some of the women doctors experienced hostility from both programme directors and fellow residents over their pregnancy.[28]

In terms of positions of power, Britton contends that while women are the major source of labour for the health industry, they are not proportionally represented among the leaders of the health services in Australia. She produces an impressive array of figures which indicate, among other things, that few women are at the top, even in social work which is a female-dominated area (77 per cent women). Surveys of the gross salaries of administration, nursing, para-medical and medical staff in Victorian hospitals indicate that in each of these groups, males have a higher average gross salary.[29]

Curran points out that in Britain, although women make up 10 per cent of people with scientific knowledge, they are less than 2 per cent of 'managers', practising research scientists and engineers. The largest proportion (36 per cent) of qualified women are teachers because of limited opportunities and patriarchal exclusion.[30]

In her book comparing the career development of women and men doctors in the United States between 1940 and 1979, Judith Lorber found similar patterns. Women were mainly in the lower or middle levels of the career structure, had problems dealing with domestic responsibilities and had often sacrificed their potential careers for the work needs of husbands. As with other highly educated women, 'they marry later and less frequently than male physicians and men in similar professions, they divorce more often, and they have fewer children'.[31]

The overall picture is that women's participation in science is growing, but is still markedly lower than that of men. Marriage and children are experienced by women scientists as a hindrance to their work pattern. Women scientists have the less prestigious jobs and are paid lower salaries. In the 'harder' sciences, for example engineering, women virtually disappear from the statistics.

Why so few? Obstacles for young women

The reasons given for the low female participation are twofold: women are not *able* to do science, because they are biologically and intellectually incapable; and they do not *want* to do science and have a lower achievement drive. Feminists have argued that these are fallacious beliefs, and that there are three reasons for the current situation: the lack of role models in maths and science for women; the education system and socialization processes which control women's 'choices'; and the nature and structure of science itself which often leads to women selecting themselves out or being selected out. When these issues are explored, the question rephrases itself: how do so *many* women manage to do science?

Rupert Hughes has commented that 'women's intuition is the result of millions of years of not thinking', highlighting the sexist rationale for the lack of women in science.[32] Women are defined as non-thinkers, subjective and body-based.

The evidence for this viewpoint has been based on sex differences in abilities and personality. However, as with much of the research into sex differences, the methodology in early studies was biased towards males. The most comprehensive analysis of the sex difference research was carried out in the 1970s by Maccoby and Jacklin, who examined over 2000 articles and papers on the subject, most of them written after 1966, finding few sex differences.[33] This work has since been criticized as lacking critical depth.[34] Debate still rages over sex differences, but a brief survey of the findings to date indicates the following.

Girls have better verbal ability than boys. In mathematical ability, boys and girls are similar in skill in pre-puberty years. Men have been found to be better at reasoning and problem-solving, but this often depends on the content of the problem and whether it is sex-based. Frieze and her colleagues give two examples of such problems—one with cement and one using cake making—to show that the difference in presentation of a problem affects the chances of solving the problem for boys and girls. The resulting sex differences may thus be the result of the different skills and experiences of girls and boys rather than differences in reasoning ability. They further cite studies which show that girls do better at maths with a female instructor—but in fact maths teachers are mainly male.[35] Most authors conclude that mathematical ability only begins to differ between the sexes when girls reach puberty and adolescence, when social pressure exerts itself and to be popular and feminine becomes important. It is at this point that girls begin to fall behind in maths skills.

One of the reasons advanced for girls' lack of success is that they have a different cognitive style or way of thinking that is not analytic. Analytic ability is the ability 'to perceive an item as separate from its background and to break, set or restructure the situation'[36] and boys are said to be better at this. The best-known test on which these statements are made is the 'rod and frame' test in which people have been asked to clarify when a rod is upright in a frame which is tilted.

These studies have found that boys are field-independent, that is, they can identify the position of the rod, regardless of the position of the frame. This is said to indicate boys' greater analytic ability and therefore suitability for science. Girls have been found to rely more on the links between the two and have been called 'field-dependent', the language reinforcing the idea that 'different' equals 'inferior'. The term 'field-sensitive' better describes this aspect of their perception. Maccoby and Jacklin also found that in non-spatial tests, when it was required that people ignore an *irrelevant* context or restructure the elements of a problem, women often perform as well as or better than men.[37]

There are a number of points worth noting about the sex difference literature.[38] The tests used are limited and help to determine the kind of results obtained. The results are always discussed in terms of averages, and many women and men differ from the average. Because there are no marked sex differences in abilities before adolescence, the influence of socialization cannot be ignored. Parents play an important role in encouraging daughters to be exploratory and adventurous. Close control and restriction of the child seems to develop field-sensitivity, while encouraging initiative and exploration creates independence and analytic thinking.[39] In terms of experience, it is clear that the daily environment of girls and boys differs. Robertson discussed an American study of students enrolled in a science-enrichment programme and found that more men had science-related hobbies than women (72 per cent compared to 37 per cent) and more of the males had a science laboratory at home (35 per cent compared to 9 per cent).[40]

Sex differences are multi-determined, and after her analysis, Sayers comments that little has changed since Helen Thompson-Woolley said of the research into sex differences in 1910:

There is perhaps no field aspiring to be scientific where flagrant personal bias, logic martyred in the cause of supporting a prejudice, unfounded assertions, and even sentimental rot and drivel, have run riot to such an extent as here.[41]

As well as being judged deficient in analytical ability, women are deemed not to have the appropriate personalities for science. Competitiveness is seen as necessary in science, as are independence, self-reliance and an inquiring mind. All of these characteristics have been attributed to men. Roe found that four characteristics are most typical of outstanding natural scientists: high intellectual ability; persistence in work and a channelling of energy; extreme independence and the ability to work alone; and an apartness from others.[42]

There are two assumptions here. The first is that women lack and cannot develop these characteristics, and the second is that science needs and should demand these characteristics. Studies of women in science are interesting in these respects. In a study of women mathematicians, Helson found that a non-conformist, individualistic view of the world, high flexibility, and originality were characteristics of creative women mathematicians. She compared creative mathematicians with other women mathematicians and found them to be rebelliously independent with a rejection of outside influence and to find self-expression and self-gratification in research activity.[43]

Bachtold and Werner studied successful women biologists, microbiologists, chemists and bio-chemists. They found in general that the women scientists were more serious, less sensitive, more confident, more dominant, and bolder than women in general. They were also more intelligent, aloof, radical and less group-dependent. They revealed themselves as people who like to work alone, are precise, deal well with abstractions, are self-sufficient, dependable and assertive.[44]

Comparing the women with men scientists, these researchers found the 'most striking impression is the similarity' in personality. They comment that: 'In some, the salient traits that had been previously reported to be characteristic of the male scientist of note—intense channelling of energy, self-reliance and dominance, a low interest in social interaction, and a willingness to take risks—were all characteristic of the successful woman scientist as well'.[45] It is clear from these studies that women can be 'good' scientists and have the personality characteristics and skills to do so, *and* that 'science', as currently constructed, places demands on personality that we should question and investigate.

Do scientists have to be insensitive to group influence, introspective and unwilling to interact socially? When discussing women, physics, and femininity, Walberg studied the difference between girls and boys in science and finding few, commented that 'girls score higher on aesthetic values and also rated the universe as more beautiful. These findings are significant for the present study only if they appear to reflect the external and perhaps *more superficial* orientations of girls' (emphasis added).[46] There is an assumption that all forms of sensitivity must be eliminated from the 'true' scientist. I will return to this point later in the chapter when discussing whether women should seek increased involvement in science and whether science should select alienating characteristics in its scientists when co-operation and social responsibility would seem to be an urgent necessity.

The second rationale given for women's lack of representation in science is that they are low achievers; that is, not only is it impossible for women to be scientific, but they have no desire to be involved because they do not value achievement highly. The label 'achievement motivation' has been developed by psychologists to define the 'general personality disposition to strive for success in any situation where standards of excellence are applied'.[47] As indicated earlier, the theory of achievement motivation was developed only with male subjects, as women did not fit the preconceived theory of the researchers. Because the specific instructions used in the studies did not arouse a need to achieve in women which was the same as that in men, it was assumed that the difference was obvious—women do not feel the need to achieve. Thus in the classic work in this field by Atkinson, women are mentioned only once in a single footnote and in McClelland's book *The Achieving Society* women are not mentioned at all.[48] The women's movement generated interest in this obvious injustice and recent work has shown the complexity of the concept.

Horner developed a theory of the fear of success to explain women's general lack in this area. Her studies showed that women were fearful that success would make them less feminine and therefore less attractive.[49] However, more recent research indicates that women strive less strongly for success when the result will be negative and they will be punished for being successful. This punishment comes in a variety of forms and can include: verbal abuse or harassment; innuendo about the woman's success as a woman; sexual harassment; the attributing of her success to luck rather than skill; and having to pay for success by being superwoman and carrying the domestic load in addition to outside work. Devaluation and rejection are effective ways of attacking someone else's identity and these are used in particular against the woman who succeeds or is competent in a 'man's world'.[50] Furthermore, Joseph Pleck indicates that men fear women's competence, particularly competence in their partners.[51]

The area is still rife with controversy, but studies indicate that in general, women see themselves as being less successful *as women* if they do well in male-dominated tasks or jobs. This negative self-image is constantly reinforced in order to ensure that women will not attain powerful positions and will not compete with men in the more powerful jobs. Furthermore, the image of women as having stronger needs for affiliation, that is, as needing to relate to people and gain their approval, means that women and their self-images are controlled by those who can give or withhold approval. In patriarchal society, 'those' are men. The so-called affiliation need also allows women to be relegated to child-rearing and home-making.

It seems clear then that far from being trapped by their biology into being under-achievers in science, women are under-represented in the scientific establishment because it is structured in such a way as to exclude and undermine them. There are other dimensions to the issue. One is that women have not entered science because the education system is structured to exclude them; because there is a paucity of known role models for girls to emulate; and because science itself is structured in such a way as to make the entry of women difficult.

In general, the school experience determines a person's future occupational path. Schools are, however, structured around patriarchal values; in a profession dominated by women, the vast majority of power positions are held by men. For example, in New South Wales in 1987, although women constituted 57 per cent of the teaching force only 8 per cent of secondary school principals and 5 per cent of primary school principals were women. (However, through affirmative action programmes the Education Department aims to fill at least 40 per cent of senior positions with women in the next five years.)[52] Leder has indicated that teachers respond differently to boys and girls in the classroom. They have assumptions about what is 'appropriate' for girls to learn. Because of the differences in treatment, boys become more confident about their capabilities and 'learn that if they don't succeed it is because they haven't tried hard enough, while girls are made to believe that failure is indicative of lack of ability'.[53]

It has also been shown in classroom studies that there are different teacher-initiated contacts for boys and girls. For example, differences between the treatment of girls and boys have been found in the opportunities they were given to respond; in open questioning; in the level and persistence of questioning; in praise and criticism, and in individual treatment. These were positive in the boys' direction. Firkin warns that 'observational studies have shown that there are very significant but largely unrecognized differences in the way in which female and male students receive attention in class'.[54] It is clear that schools reinforce social values to the advantage of men.

Concern is focused at present on the proportions of girls in maths and science for good reason. Girls are cutting off their options in these subjects too early in their schooling. As Kelly and Weinreich-Haste comment:

Girls who cannot or will not learn science are cut off at an early age from a wide range of careers and interests. By conforming to the feminine stereotype which excludes science, they are moving towards traditional women's occupations, and the low pay and low status which frequently accompanies such occupations.[55]

Traditional jobs for women are also the ones being hit hardest by new technology: copy typists are being replaced by word processors; sales assistants by point-of-sale computerized terminals; bank tellers by automatic cash dispensers. Yet the majority of women in the workforce work in clerical, sales and service positions.

Attitudes to science crystallize in the early teens, so the experience girls have at this point is essential. But few teachers at high school level, and fewer at university level, are women. There is a need for girls to be shown women already successful within scientific fields. In general boys develop a more favourable attitude to science than do girls. Kelly comments on this with respect to the United Kingdom, where girls have been found to view science as hard, complex and dull, and concerned with things rather than people. This view is shared by boys, but it does not conflict with their self-perception and the role society affirms for them. Girls therefore do not do science because of its masculine image.[56]

Weinreich-Haste studied the attitudes of school students in Great Britain to various subjects. Physics, chemistry and maths were seen as masculine by both sexes. Boys considered history and biology to be masculine subjects, but girls considered them neutral. In total, girls saw two subjects as feminine, two as neutral and three as masculine; while boys saw two as neutral, five as masculine and none as feminine.[57] In an Australian study of grade 4 girls and boys who were learning to use Apple computers, Chambers and Clarke found that girls are apprehensive or fearful of computers and stand back to allow boys access. They feel boys have greater access to the computers. One hundred per cent of girls said that boys spent more time on the computers than did girls and only 15 per cent of boys thought so. Girls also felt that their friends did not want them to learn about computers.[58]

Science is often seen as dull and unconnected with people. Alison Kelly cites a study which showed that girls are more likely to study science if they consider that it has relevance to their lives. She suggests, as do others, that the teaching strategies needed for teaching girls science may differ from those used for mainly male classes.[59] Too often a discussion of the social implications of science is restricted to the final year, if it ever occurs, *after* girls have opted out of science.

Ingvarson also stresses the alienation effects of the way science is taught and the assumptions that students will be male, stressing experiences which boys are familiar with, for example, engines or cricket ball trajectories. She lists many useful ways to improve the classroom situation and stresses the need for girls to meet working women scientists and mathematicians.[60] Role models are essential for girls. History, like most fields of knowledge, has been male-dominated and women who were successful in maths and sciences have disappeared from view; Marie Curie is often the only known woman scientist. Because of the structures of science (discussed below), there have been fewer eminent women than men. But a knowledge of those who have been eminent acts to make women 'normal', a part of the human enterprise, and capable of succeeding in these fields if opportunities are available.

Barriers within science

Zuckerman and Cole have delineated three barriers to women becoming productive scientists: the fact that 'science is culturally defined as an inappropriate career for women'; a belief that women are less competent than men; and actual discrimination against women within science itself.[61]

Discrimination within the sciences takes a variety of forms. The social isolation experienced by women in science is dominant. Male colleagues avoid them and find it difficult to work with them because women are not supposed to understand science and it makes men nervous when they do. Women find themselves either abused, ignored, or used.

Reskin has clearly explored the social models open to women in science. She notes that the collegiate role is stressed, in which scientists 'teach, collaborate with, encourage, inform, evaluate and reward, compete with and befriend co-workers'.[62] Collegiate co-operation is based on competition with the right people and co-operation within your own team or university. But sex differentiation blocks access to this collegiate relationship because of women's lower status. The models of cross-sex relationships do not include collegiate relationships.

Another model is the father—daughter model which leads to a paternalistic treatment of the woman scientist and therefore a devaluing of her. A relationship based on marital roles works for some, but often leads to a junior authorship for the woman. The model of the traditional heterosexual romantic liaison is also dangerous. Women are held to be responsible for any sexual liaisons. Those who are suspected of using their sex for advantage have doubts cast upon their professional achievements.

Yet those who eliminate all sexual and romantic aspects from their interaction pay the price of being disdained as cold and unfeminine. Being in a less powerful position in hierarchical science, the woman is also prey to the pressure of sexual alliance. It is difficult for a woman to be comfortable seeing herself as both scientist and woman. The conflict between success and desirability as a woman tends to make some women 'vulnerable to a subtle kind of exploitation by the patriarchal hierarchy'.[63] The Science for the People collective writes: 'In our experience it is surprisingly common for the male leader of a research group to enter into an extra-marital sexual liaison with a female subordinate, such as a graduate student, post-doc, or technician.'[64] They note, however, that in most cases the exploitation is more subtle, in the form of innuendo and flirtation. These are difficult for a woman to ignore, since they provide a means of lessening the aggressiveness or defensiveness of male colleagues. Because of the possibility of pressure in these directions, 'even women who would never dream of sexual involvement with a co-worker monitor their behaviour to avoid the appearance of trading on their sexual attractiveness'.[65]

Within scientific training there is a need for a sponsor, a person of power, who has the interests of the young scientist in hand and can direct and advise.[66] But because men tend not to see women as their successors in the field, they may be unwilling to sponsor a woman as a colleague.

Informal communication is essential within scientific professions. In casual conversation, an exchange of ideas and information takes place which is invaluable and often not attainable through published material. Information about new unpublished material, and who is worth knowing and who is dangerous to know, can affect the career path. Conferences are also important for this informal exchange. However, women are rarely included in these informal networks simply because as women they have lower status and are seen as 'wastage'. Reskin comments that men see women as passive receptors, the receivers of information, and so exchange on a professional level is impossible.

Patterns of collaboration are also sex-based and exclusively male. Collaborative efforts begin informally and so women are less likely to be approached. Women would also devalue the partnership or project by virtue of their gender, so few men are willing to include them.

These issues are part of the socialization into the scientific professions. Women are denied what White calls 'informal signs of belonging and recognition'.[67] Success within the scientific professions, as in other areas of the workforce, relies not simply on a matter of knowing the work involved and being competent. Of equal importance are the informal bases of power and the appropriate behaviours involved, for example how to write funding proposals, and how to get access to information which is difficult to obtain. These skills are learned from interaction with professional colleagues. If women are cut off from that, they are cut off from access to vital information within their career path.

These issues are important because it has been found that the acceptance and recognition of colleagues and their appraisals and criticisms of a scientist's work make 'a significant difference in determining whether a woman felt like a professional, and whether she in turn had a strong sense of commitment to future work'.[68] Her sense of her identity as a scientist depends upon these feelings of inclusiveness and support. Lorber's study of physicians analyses the role in terms of help or hindrance, by colleagues, patrons, mentors and, for men, wives. Whether the doctor works in the community or in a hospital, 'the pattern of advancement is still along the lines of sponsorship and patronage'.[69]

Science is structured as a male career, in that the assumption is that the scientist's first and only devotion is to the job. Full-time total commitment is demanded. But many women have devotion to the project in hand, as well as a desire to competently raise children or have other interests in their lives. Women are accused of a lack of commitment, of being unreliable and unstable, because men expect their career patterns and motivations to be the same as those of men. Science itself is thus structured to force choices upon women which are not conducive to their fulfilment as a whole person.

The embodiment of women's exclusion from science is the story of Rosalind Franklin, though such cases abound. Rosalind Franklin was a talented and productive young physical scientist. Educated at Cambridge, she worked first in England, and then in 1947 moved to Paris, where for four years she studied X-ray diffraction and worked on the structure of carbons. In 1951 she applied for and received a Turner—Newell Fellowship in order to join Professor Randall's unit at King's College, London. Letters from Randall attest that she was to start developing an X-ray camera to study the structure of DNA.

Also studying DNA, but at Cambridge, were Francis Crick and James Watson and at King's College, London, Wilkins and Gosling (a PhD student). In 1962, Watson, Crick and Wilkins were awarded the Nobel Prize for their contributions to understanding the structure of DNA. Watson wrote a best-seller on the discovery titled *The Double Helix*. However, the book shows Franklin in a very unfavourable light, and Anne Sayre has written *Rosalind Franklin and DNA* in defence of her. Sayre's book highlights the issues discussed above. The sexual politics of the laboratory setting become clear.[70]

Watson's own book unwittingly depicts the men as chauvinistic, vicious and deceitful. In his book, Franklin is discussed as 'Rosy', though always known in reality as 'Rosalind'. This, together with Watson's description of her physical appearance as unfeminine, belittles Franklin's character and personhood. As Sayre notes, Watson presents her as 'the perfect, unadulterated stereotype of the unattractive, dowdy, rigid, aggressive, overbearing, steely, "unfeminine", blue-stocking, the female grotesque we have all been taught either to fear or despise'.[71] She was clearly too 'uppity' for the men.

Watson further manages to turn Franklin into Wilkins' unco-operative research assistant, when in fact she was of equal status, but higher publication rank. Sayre consistently attacks each element of Watson's picture of Franklin, showing each to be either false, or a misrepresentation. The reason for Watson's personal attack on Franklin may lie in the use of Franklin's data on DNA without her knowledge or consent, by Watson and Crick. By portraying her as deserving of contempt, they managed to expurgate their own sins. Franklin, had she lived (she died in 1958 aged 37), should have been a strong contender for the Nobel Prize, which is not split more than three ways, and not awarded posthumously. But Franklin's work and con-tribution has repeatedly been ignored or discredited.[72]

In 1952, Franklin made arrangements to leave King's College and go to Birkbeck College. This necessitated her ceasing work on DNA. She thus abandoned a problem on which she was making considerable progress because of her acute discomfort at King's. 'Her loneliness and isolation within the English scientific community, her difficulties with her male colleagues, the denigration of her work and professional status by those same colleagues, and finally, the appropriation of her work by Watson and Crick', exemplify male dominance in science and its structures.[73] Hubbard comments that the treatment of Franklin in Watson's book is *not* idiosyncratic, or unusual, but a reflection of the attitudes of male scientists.[74] As Richards and Crossley comment:

The moral of the Rosalind Franklin story is that, contrary to popular expec-tations (and many academic studies!) no amount of individual motivation, ability and dedication, or alleviation of domestic responsibilities can en-sure professional success and recognition in science for women, when the very structure of science functions discriminatorily against its women participants.[75]

Women resist science as definitive and exclusive

Women's attempts to infiltrate science represent a rejection of the limited definition of woman, and a demand to be acknowledged as intellectual and rational beings. The history of this resistance has been conveniently erased and overlooked in man's push to define woman by her exclusion.

The intellectual or scientific woman has been defined by men as an oddity, the unusual or special case. But women have often led other women in their intellectual pursuits and have often played important roles within science.

In her book *Hypatia's Heritage* Margaret Alic has traced the history of women scientists from antiquity to the late nineteenth century.[76] Documenting the fascinating work of women in astronomy, chemistry, physics, mathematics, medicine and science in general, she points out that women were often recognised and respected scientists in their time, but later historians discredited or ignored their work, and importantly, their role in scientific communities. Many of these women also wrote under pseudonyms and published their work anonymously in order to protect their positions. Many of them worked with famous male scientists who have often been credited with the women's ideas. Mostly they were privileged women, and in times of lack of access to formal education for women, managed to educate themselves through independent means or through the help of fathers, brothers and husbands. The book ends before Marie Curie's discovery, at which point science had ceased to be available to the amateur and was becoming a male-controlled profession.

Sally Gregory Kohlstedt also traced women involved in science from 1830 to 1880 in America. She documents the achievements of botanists and naturalists or women who wrote chemistry textbooks like Almira Phelps; and traces the opening up of the sciences by women until they began to challenge male associations for entrance. Kohlstedt concludes:

The hopes and triumphs of women in the 1880s indicates that they were a 'first generation' of public achievers. Their success, however, was built on the persistent efforts of women who had worked quietly in science throughout the century, women who both reflected and expanded contemporary expectations. The female illustrators and textbook writers had not challenged the boundaries of women's sphere, but their works refuted glib assertions about women's capability to understand science ... most important, by their own example, the forerunners demonstrated to those who chose to notice the capabilities of women in science, and they established co-operative networks, which would be essential for expanding women's opportunities in science.[77]

In England, Mary Somerville (1780-1872) established herself as a mathematician and in her lifetime published a number of books in her field, the last when she was 89 years old—*Molecular and Microscopic Science* (1869). She published on mathematics and physics, and received considerable honours in her time. She was self-taught in her field, education not being open to women, and was aided by the death of her first husband which left her independent of means and able to pursue the studies her husband had disapproved of. Her second husband supported her work, though she complained of constant interruptions from the children! The Somerville College at Oxford, and the Mary Somerville Mathematics Scholarship for women at Oxford remain in memory of her creative talents. But few will know her name and 'she too has virtually disappeared, and so extensive has the repudiation of her been that it is still possible for people to assert that women are prevented from participating fully in mathematics and science because of some defect in their makeup'.[78]

Australia too has its heritage of invisible foresisters, though there is much yet to be documented. O'Neil has outlined the history of women at Sydney University who were admitted to degree courses in 1881. Dagmar Berne entered medicine in 1885, and after four years, completed her work in the United Kingdom where she could not have obtained a degree, but where she could obtain a hospital residency, leading to registration. She returned to Australia and practised in Macquarie Street, Sydney. O'Neil names a number of women who gained medical degrees after 1893, but notes that they all found it impossible to obtain a residency which was required for registration in a Sydney hospital.[79] In the early days of Sydney University, male students threw bits of dissected cadavers at women medical students.[80] In 1865, the *Medical Journal of Australia* supported the Medical Board's refusal to register the first woman doctor (Winifred Ferguson) who came to Australia. It stated:

If Winifred Ferguson favourably impressed our legal experts, she produced quite the opposite effect on the medical profession . . . But there is little to fear that in any British community medical women will exist as a class. They will occasionally be imported like other curiosities, and the public will wonder at them, just as it wonders at dancing dogs, fat boys and bearded ladies, and, in accordance with the demand for novelties, they will, perhaps, be as successful in a material sense, but they are not likely to be included in the list of British Institutions.[81]

Dr Kate Campbell, who started medicine at Melbourne University in 1917, recalled that there were twenty-six girls in a class of 160 and that this 'high' number was because during the war women's opportunities broadened. She recalls the sense of shared experiences and interest the women had and that 'sometimes a lecturer would embarrass us with coarse jokes'.[82] When doing clinical work, the men were assigned to top clinicians, but the women's first outpatients' clinic was the male VD clinic. When given their residencies at the Melbourne Hospital (the Alfred and others were closed to women) they were given the dirty work to do, and were not allowed the invaluable experience of casualty wards. Hospitals used a lack of toilet facilities as a reason for excluding the women.

In 1896 Queen Victoria Hospital in Melbourne was established by twelve women who wanted 'a hospital where women could be treated by women doctors, and where women doctors could treat patients'. Campbell notes that the Queen Victoria became the cornerstone for women in medicine. She recalls: 'I remember one woman doctor telling me that her medical course had taught her how to treat a patient as a doctor; the Queen Victoria taught her how to treat a patient as a woman'.[83] In 1956 male residents were admitted and the staff gradually became male-dominated. It is ironic that the leading test-tube baby team worked here, reinforcing the value of women as mothers only and using women's bodies as living laboratories.[84]

Unmarried, Campbell talked of the women forced to give up medicine by their husbands, or refusing to marry because the cost involved was the dropping of a medical career. She noted that women's position has improved greatly now due to their earning power and therefore independence. 'The inferiority of women was due to three "Ms": muscle, maternity and money'.[85]

These women in science, medicine and mathematics can form important models for other women and girls. They show that women can succeed in these fields and that they do have the abilities necessary. They also indicate the strength of women to achieve in spite of the odds against them.

In order to rectify the lack of information given to girls on these foresisters, the Equal Opportunity Unit of the Victorian Education Department produced an impressive kit on 'Women in Mathematics and Science' which traces the experiences of nine famous women scientists and nine well-known women mathematicians. In both groups, the women were active researchers, writers of textbooks and teachers, and three have won Nobel Prizes in their fields, Marie Curie twice. Thanks to constant pressure in this area, the CSIRO has established a programme of visits to schools by women working in scientific research and technology. They talk to middle school girls with the hope of encouraging them into science.[86] In addition, women scientists have formed a network to support each other and an association to publicize their work, and to change science and its values. *Wisenet*, the publication of the group, keeps women up to date on issues for women in science.

The battle to keep women out of science also continues. In 1972, headlines in Sydney newspapers ran: 'Limit women: surgeons' plan' and 'Surgeons want women cut out'.[87] More recently the Federal government has introduced affirmative action legislation applicable to tertiary education institutions. In response, nineteen Deans of Science (all male) met the Minister for Education and the Minister assisting the Prime Minister on the Status of Women to protest this imposition.[88]

A study of our foresisters in science revives again the issues of marriage and children in the life of a woman scientist, and the difficulties involved in juggling these and their career. Many women find the choice an either/or one and opt for the career alone, or vice versa. Some women scientists continue to be presented as 'superwomen', managing both worlds. However, this representation perpetuates the unspoken assumption that *all* women who are interested in science can do this too—or there is something wrong with them. An article on Marjorie Rhodes Townsend, a NASA Satellite Project Manager exemplified this. Townsend has shown 'outstanding technical and managerial leadership' and been awarded NASA's Exceptional Service Medal. But, 'she also found time to bear four sons. Working with a woman who could have four babies without ever having to take leave without pay was quite an education for her bosses'.[89]

In a panel discussion Sichel noted: 'Admiring interviews in newspapers and magazines suggested that every successful woman scientist had five lovely children, was a terrific cook and housekeeper and had a myriad of other talents. In fact, this may be true, but it certainly discourages most women from seeking an active scientific career'.[90]

Many women scientists themselves claim that it is the nature and structure of science itself which acts against the inclusion of women. The attitude of men within these professions has ensured that they remain male preserves both with regard to structure and to content.

Should women be in science?

Well, should women be striving to enter the sciences? The affirmative position argues for more women in science. The more there are, the harder it is for men to argue a biological predetermination for their exclusion. Women already within science also change their perceptions of themselves if they have support. How much easier it would have been for Rosalind Franklin if there had been another one or two women scientists working with her. Curran has said that 'one of the reasons women don't want to study science or work in scientific jobs is because it is unpleasant to be in such a small minority'.[91] She quotes one woman scientist as explaining the subtle prejudice used against women, and the effect of its daily occurrence which 'drains your self-confidence and motivation'.

Women scientists also find the hierarchical structure and competitive environment hostile. They comment on the undesirability of acquiring the necessary 'male' values needed for success. Here 'openness is anathema', 'emotional dishonesty is blatant under guises of reason, objectivity and abstractions', and 'the social reasons for doing science are lost among the emotional needs of Western man to achieve, perform and acquire status in the eyes of their own sex'.[92] Some women feel that against this, numerical strength helps!

There is also the real thrill of discovery for the scientist and the belief that they are contributing to the overall good. Women scientists enjoy the work itself and find it exhilarating. In addition, Wu has argued that women's perspective is needed in science:

Men have always dominated the fields of science and technology. Look what an environmental mess we are in. They have brought us to the gigantic brink of environmental ruin. The air is polluted; lakes, rivers, seas and oceans are contaminated. Women's vision and human concern may be exactly what is needed in our society.[93]

The so-called ability of males to isolate one aspect and to narrowly pursue it, which has been presented as necessary for the creative scientist, can be questioned in the light of the lack of social understanding shown by many male scientists. If women's skills lie in considering context and being 'field-sensitive' perhaps these are the necessary skills needed to ensure a more wholistic outlook for science. Hubbard describes this position as being opposite to that of Watson and Crick's. It is one where the scientist sees things in context and does not strive to isolate events or things for study. 'It is not', Hubbard writes, 'pessimistic to recognise that in the real world everything is connected with everything else, that "isolated variables" are figments of a mis-educated imagination'.[94]

Women need to enter the sciences also in order to monitor and change the control of women through science. Because science strives to control nature and dominate it, the threat exists that women's natures and bodies can be controlled.[95] The medical stranglehold on pregnancy and birth is evidence of this. The new reproductive technologies are another example. Control of information like this becomes a power base. Because women do not control these spheres, they often become experimental guinea pigs in the scientific enterprise.

Furthermore, male dominance in the profession of medicine creates a male way of viewing the female patient. Advertisements for drug companies have been shown to be stereotypical, reinforcing a male-dominated view of women and their problems. Stimson quotes an example of an advertisement for a combined tranquilliser and anti-depressant. It had a picture of a woman in a crowded living room with five children. The caption said: 'Lack of space, lack of privacy, spawn unhappy people'. The solution was to prescribe a drug.[96]

The development of nuclear power is another example of the anti-human nature of science and technology, where for the small amount of power utilized, we are faced with 'an insurmountable nuclear waste problem, a health risk, the possibility of serious accident'. Part of the problem as Thompson sees it, is that women have no part in designing and working science and technology.[97]

Science is changing society, and unless women are involved in science, all the changes will be directed by men for men. Women need to be involved in current debates on, for example, pollution, and need to be informed to be able to do so. Sandra Harding has argued that affirmative action strategies in science can be seen as both reformist and revolutionary, 'primarily because desirable directions for radical change emerge only through our attempts to make what one might have thought were merely reforms, and because "mere reforms" have nevertheless created resources for those radical changes'.[98]

If a woman does have this understanding, she will have a feeling of competence and mastery. Even being unable to carry out simple repairs to machinery which daily surround us, lead women to rely on male experts. Women need to feel competent in an increasingly technical and mechanized world, and this feeling of competence will lead them to have a sense of personal control, essential to a strong self-identity.

In spite of these strong reasons for women's infiltration of the sciences, many feminists argue that this is not a real solution to the problems of science. The main problem is that science is masculine and has all the negative values of masculinity which feminism fights against: exploitation of people; destruction of nature; and control and dominance over people and nature. Science has become a tool of patriarchy with its legitimation of the subordination of women. The deterministic justifications for women's role have bloomed from science. Overfield expresses this clearly: 'The scientific ethic as much as the capitalist and imperialist ethics, was based on exploitation, elimination of rivals, domination and oppression'.[99]

The language of science itself places the world against itself, dichotomizing it: objectivity is reified and subjectivity scorned; engineering parts are masculine if they penetrate, feminine if they are passively penetrated.

It is also argued that women who enter science are required to become 'honorary' males: to assume the competitive, oppressive values of science. The socialization processes of science itself ensure that these values are difficult to resist. More women in science will not in itself change this, nor the corporate and commercial links medicine now has.

Furthermore, science has become part of the productive mode of capitalism, in which the end product rather than its impact in human terms, is the goal. The values of science have lost contact with the values of humanity. Over and over we ask why people starve, why women cannot have a safe contraceptive, why breast cancer is still raging, while men put themselves on the moon or into orbit, thus symbolically distancing themselves from the concerns of earth.

Wanting power over others is incompatible with caring for their *real* needs; oppressors become selfish and contemptuous of the oppressed. They lose their understanding and empathy. The real dichotomy is life versus death. And it is in the direction of death and destruction that science is moving. The links to male power and sexual dominance become clear in the American term used for what male scientists see as the inevitable nuclear war: 'wargasm'. Thus, Stehelin has written that 'fetishism, production, science, are all linked according to the phallic code—"the no-penis is no-knowledge"'.[100]

So, there are two options available: to encourage more women to enter science, and to change its patriarchal structures. Science itself is in crisis because of over-specialization. Scientists themselves can no longer see past their narrow enclave of specialist knowledge or see the relevance of much of their work. Fausto-Sterling suggests that women must enter science and change its boundaries; break down the hierarchical structure; make the content more socially relevant and change the language of science. Women must move it away from the abstraction which has been seen as the male province, and make science accountable to people; particularly to women, who are so often its victims.[101]

Bleier also argues for changing the nature of science. The dualistic mode of thought should be dispensed with, as it is false in the face of the constant change, flux, and interaction which life is. Science needs a sense of context and inclusiveness for its work, not isolation and exclusiveness.[102] Science should acknowledge its value base and be self-critical. And it should struggle to be relational and accountable in the world.

Science has been used by man as a tool for defining women as part of nature, not culture. The content of science has reinforced patriarchy, while posing as established truth. It has defined women by exclusion from science, attempting to reinforce negative self-definition. Women's resistance to both the self-definition and the exclusion has exemplified the falsehoods within science itself.

3
Culture and Self-creation through Writing

Just as women have been both defined out of rationality by science and then excluded from science by its own structures, they have been defined negatively within language and literature—and structured out of its development. Even the emotions and sensitivities which women are said to possess and which problematize their rationality have not ensured them a place in the world of creativity, but rather have excluded them once again on the grounds of a lack of *disciplined* emotion.

The exclusion of women has meant that cultural representations of women, whether through film, television, visual arts, theatre, music, dance or writing, have been male-defined: woman once more 'created' by man, in relation to men, and for the purposes of men. This chapter will deal with some aspects of literature and language itself as a case study in the way cultural representation is used to reinforce the male definition of 'woman', and the way men try to control women by excluding them from the culture-creation process. It will deal with the expression of women's resistance through their writing.

Literature as a tool of self-definition

Literature (and language) is a powerful tool by which the consciousness of a society is controlled and reinforced. 'Through it', wrote Charlotte Perkins Gilman in 1911, 'we know the past, govern the present, and influence the future ... The makers of books are the makers of thoughts and feelings for the people in general'.[1] But it is difficult for a person to distance herself from the dominant culture in which she lives. Literature and the arts fall within the male precinct of 'culture', which is set in opposition to 'nature'. Women have been the victims of, but also the supporters and consumers of, male culture.

Through language our stories are made and told. And these stories help to explain ourselves to ourselves; to understand the human condition and to see ourselves in the experiences of others. But this picture has been skewed. For most of history men have told the stories. The very act of creative writing is seen as masculine, because it requires control, self-discipline—and leisure. It is here, within culture, that men attain their control over creativity reproduction. Women may be able to create life, but men create culture. So Gerard Manley Hopkins wrote in a letter to his friend, R.W. Dixon, in 1886: the artist's 'most essential quality' is 'masterly execution, which is a kind of male gift, and especially marks off men from women, the begetting of one's thought on paper, on verse, or whatever the matter is . . . The male quality is the creative gift'.[2] This is an earlier version of the now famous comment by Norman Mailer that the prerequisite for good writing is 'balls'.

Women have been asked to identify with the male picture of woman: dichotomous myths woven around the dark and the light woman; whore versus madonna; angel and madwoman. For example, the images which emerged in the twentieth-century American novel, have been detailed by Rodgers.[3] Her survey demonstrates the view of woman as temptress, saint, or emasculator of men. Women writers, she points out, often also used these same images through their inability to transcend the stereotypes and the pressures of publishing. But women's use of these stereotypes often carried a different message, warning readers of the end results of pursuing stereotyped roles. Rodgers explores the dark and fair lady; the evil goddess; the violated woman; the charming betrayer, the bitch, the obsessive mother, and the corruptors—what Register has called misogynistic polarities.[4].

The choices faced by these unreal characters are those men see women as having, not the complexity of life as experienced by women. So men's stories of 'life' continue to validate each other and the experiences of men. And they exclude the reality of women's lives, for 'the world women experience is demonstrably different from the world men experience'.[5]

If this world of women is written out of literature, women cannot find themselves in an historic or contemporary continuity of experience with other women. What happens then to their self-identity? It becomes again fragmented, isolated. The mirror of the arts reflects them as object, evil, mere nature, sexually obsessive and consuming, and always in relation to men—but not to each other. Carol Christ has written:

Women's stories have not been told. And without stories there is no articulation of experience. Without stories, a woman is lost when she comes to make the important decisions of her life. She does not learn to value her struggles, to celebrate her strengths, to comprehend her pain. Without stories she is alienated from the deeper experiences of self and world that have been called spiritual or religious. She is closed in silence. The expression of woman's spiritual quest is integrally related to the telling of women's stories. If women's stories are not told, the depth of women's souls will not be known.[6]

Power relations between the sexes are inscribed within the texts so that they are unquestioned and therefore implicitly, if not explicitly, acknowledged as fixed and inevitable. From male writers, women get a clear picture, not of woman and her experiences, but of what men have thought these were or should be. Feminist writers and critics have called for a revision of male texts, what Adrienne Rich has called 'the act of looking back, of seeing with fresh eyes, of entering an old text from a new critical direction', which is an essential 'act of survival'.[7] Steeped in male representations of women, women readers can unconsciously collude, so feminists call on them to resist what Kolodny calls 'the sexist designs a text might make upon her '. These involve women, for example, in identifying against themselves, when their sympathies are manipulated or conscripted in the cause of the male hero 'but against shrew or bitch characters'.[8]

The problem is that women are required either to continually experience and identify with the literature which mythologizes them in a negative way and misrepresents them to themselves; or to identify with a male world which is not theirs. Few male authors create characters with whom women can identify and who will increase their understanding of being woman. Choosing between the 'simpering thoughtless angel' and the independent, 'ball-breaking' monster is hardly a choice. As Elaine Showalter comments:

Women are estranged from their own experience and unable to perceive its shape and *authenticity* in part because they do not see it mirrored and given resonance by literature. Instead they are expected to identify as readers with the masculine experience and perspective, which is presented as the human one.[9]

Showalter believes that women's experiences have not been fully explored or expressed by women writers either, due to the restrictions imposed by the 'double critical standard, by a double social standard, by external censorship, and, most dangerous, by self-censorship—which is sometimes exercised in self-defence, more frequently in self-hatred'.[10]

This self-hatred is reinforced in the negative representations of women within literature. In her review of *Who is She? Images of Women in Australian Fiction*, Judy Turner points out that of the six male authors dealt with, four simplify and typecast women (she exempts Patrick White and Martin Boyd). David Ireland violently negates women, Thomas Keneally sees women as the bearers of a sickness called sexuality, with bodies whose functions are often associated with evil and mystery which are punishable. Henry Lawson and Joseph Furphy put women on pedestals, 'partly because they are easier to sketch when they are sitting still', partly because of guilt feelings, and partly because of a 'dreadful blindness'.[11]

Cheri Register believes that the modern American male novelist is becoming even more misogynistic in response to rising feminist consciousness. Among Australian novelists David Ireland could well be a case: in his fiction, Elkin comments, 'the women characters tend to aggregate to one or other of two poles; that of very special girls, or that of the rough-and-tumble sheilas'. Ireland wrote novels in which women, 'though indispensable', were 'incidental and peripheral to the real business of living, which is yarning, bragging, brawling and drinking'. He decided however to make woman his central concern in *A Woman of the Future* and *City of Women*. He did this because he detected the message that 'Australian male writers didn't seem to be capable of writing convincingly about female characters'. Feeling that he could, he wrote *Woman of the Future*, and proved the critics right.

This offensive book is a documentary-style story of all that a man would imagine a girl's growing up to be concerned with—primarily sexual experiences with young men, old men, herself, and her father. Elkin is critical of it: 'He writes too much about her as a specimen and a type, and, finally, as a symbol . . . generalized and allegorized'.[12]

City of Women is more misogynistic however, with its portrayal of women as just another version of men. The book is peopled with male stereotypes bearing women's names, all of whom are unlikeable and vicious. The rapes of women by Jack the Zipper who infiltrates the city are revolting. His usual method is to knock out the woman, cut into her flesh (usually the leg) and deposit his semen there before sewing up the wound. One woman is 'brain raped' and each detail is methodically described, even to the 'strong penis, the first pressure against the soft brain tissue—the probing entry . . . the hairy pubis pressed against the brain case'.[13] This book is not about women at all, but is a misogynistic exploration in male fantasy.

The idealized stereotype of women as either madonna or long-suffering victim or, as Gilbert and Gubar put it, the 'angel' (as opposed to the 'monster') is just as anti-woman. The angel represents selfless purity and 'to be selfless is not only to be noble, it is to be dead'.[14] This idealization obscures the reality of women's experiences and their interrelationship with men and with other women.

Australian women have a particular problem in trying to establish an identity within a culture that is particularly exclusive of women apart from their roles as 'damned whores' or 'God's police'. As Delys Bird has commented, 'Australia was statistically a predominantly masculine world at its inception; ideologically it has always been one'.[15]

Strongly influencing Australian writing has been the preoccupation in male culture with an idealization of rural living and the development of the 'bush ethos'. It dominates our literary heritage. This preoccupation was an attempt by the colonists to reconcile their exile from home with the harsh reality and vastness of the Australian landscape. By introducing whimsy or myth around life on the land, the new Australians could more easily adjust to it. Life in the bush was thus eulogized and romanticized by writers such as Henry Lawson. Lawson and Paterson created narrative ballads around bush life which had a lasting influence on Australians and on poetry. They created a model which could be embraced or severely rejected, but either way, writers reacted to it and thus ironically continued to reinvigorate it. The myth was male-dominated and to this day most Australians think of the 'sunburnt bush*man*' in the myth. The images have infiltrated Australian culture.

Rodney Hall has written of the bushman tradition as 'another aspect of the invasion, another attempt to master and possess the land, rather than learn to be possessed by it'.[16] Judith Wright discusses the need of the writer to 'be at peace with his [her] landscape' before being confident about exploring human relationships. But this landscape has seemed to have a life of its own, hostile to human intruders, with unpredictable extremes of behaviour: fire, drought, flood. Rather than learn a way to integrate these characteristics into their life patterns and their writing, Australian men have fought it, seeing the land as needing to be brought into subjection, very much like the need to subdue women as nature.

The battle between man and land for survival is the stuff of men's romantic heroic myths. The experience of the woman in the bush was very different. Isolated from company or from legal protection, she faced the dangers of bush life and the male world, represented by her husband and other men who roamed freely in and out of Lawson's bush stories.

Miles Franklin, while colluding in nationalistic myth-making, also represents the dilemma of women writers in this respect when in *My Brilliant Career* and *All That Swagger* she represents the drudgery of a woman's lot in the bush and the gradual wearing away of her health. A recently rediscovered article by Louisa Lawson also represents the lot of the bush woman, who is 'thin, wiry, flat-chested and sunburned' because of her limited diet and hard life. She 'works harder than any man' and her life 'may be nothing but a record of ill-usage' though she remains silent on this. And the feminist Louisa saw why she suffered in silence: because 'she thinks her lot peculiar to herself'.

Louisa then recites stories of male violence and brutality in the bush: the wife whose husband let her ride alone to the midwife and who, falling ill on the way, died and was eaten by native dogs; or the woman who hid three days up a tree 'while her husband was hunting for her to "hammer her"'. She saw the vulnerability of the woman who was worked or bullied to death and 'must bear what ills he chooses to put upon her, and her helplessness in his hands only seems to educe the beast in him'.[17] Her hopes lie in the daughters of these bush women, who represent strength and fearlessness as well as education and enlightenment and who will be worthy of 'women's rights'.

This article is interestingly compared with the work of her famous son the poet Henry Lawson by Zinkhan, who shows that both deal with the fear of childbirth, the masculine work of the women, the absent husband, the fear of madness in the bush, and that there are similarities in the physical description of the women. Many of Henry's stories seem to be based on Louisa's oral accounts of her experiences in the bush. She did, after all, give birth to Henry in a tent on the goldfields and lived that constantly-moving bush life for many years. Before the publication of Louisa's article on the Australian bush woman, Henry had hardly dealt with her. Zinkhan indicates that not only in choice of subject could Louisa have influenced Henry, but in his use of language also, particularly in 'The Drover's Wife' which has passages resembling Louisa's 'The Australian Bush-woman'.[18]

The most interesting point Zinkhan makes, however, regards the treatment of human menace, compared with that of the bush itself. Here Louisa and Henry differ. The protagonist of 'The Drover's Wife' fears the 'outsider' male while Louisa's bush woman fears the 'insider'— her husband. Similarly, in an interesting attempt to use Elaine Showalter's theory of the Female, Feminine and Feminist stages of women's writing (see below), Frances McInherny has considered Miles Franklins' *My Brilliant Career*. Here again, the harsh natural environment is seen as less of a threat to women than 'the male characters who inhabit that environment and mistreat, abuse, rape and sometimes kill their female "companions"'.[19] The humiliation and lack of self-esteem in Sybylla, the main character, is seen as representative of women's condition in Australia. McInherny points out that Sybylla experiencing herself as the Other is frightening and poignant as, for example, she goes through the self-denigrating morning ritual of mirror-gazing while telling herself of her ugliness and unworthiness.

As a result of what has been determined as appropriate or inappropriate for women to write about, women have often had to clothe their message in the outer garments of respectability. They have been troubled by the need to explore their own truth of the world, while concealing it (see below for a little more detail on this.) So feminist critiques urge readers to seek behind the evasions and concealments to find the hidden plot or story which is 'in some senses a story of the woman writer's quest for her own story; it is the story, in other words, of the woman's quest for self-definition'.[20] Carol Megibow also sees this as an essential issue, noting that a woman's attitude to the 'mythic tales of the patriarchy' will affect both the content and form of her writing. She sees women who harness the mythic guise themselves as strongest. She writes about women's evasion: 'women are masters of subversion, having learned the act as a tool for survival in patriarchy'.[21]

Twentieth-century women writers find it easier to represent women's worlds in their art, refusing the subterfuge necessary in the last century. They no longer conceal their emotions and have therefore shocked the literary establishment with their anger, rage, and retaliatory trivialization of the literary canon. Erica Jong's poem 'Literature Need Not' exemplifies this:

The temple of art is a locked restroom.
Only those with high class bladders

may relieve themselves within.
For the rest of us—
Women, blacks, third-world pissers,
Jews who refuse to be tamed—
the restroom is barred.
Ah! We will take our leaks outside—
in the open air.[22]

Language in the construction of patriarchy

Representations of women in male writings reflect a misogynistic and limited picture of how men see women. But language itself, the tool of writing and speaking, is also male-dominated. Words, said Aldous Huxley, 'are matters of the profoundest ethical significance to every human being'.[23] They are also of political significance. Language is not a 'given' but a created means of communication. Our use of language determines to an extent our experience of the world. In psychological terms it gives the mind a 'cognitive set', that is a pattern through which to perceive reality.

Women and men have a different relationship to language because men have made themselves the controllers of language and the 'namers', the legitimators. They have decided what is to be labelled and how. New terms have been coined until recently only by these power-holders. Language has been yet another male club, not in a spatial sense, but in a psychological and symbolic sense. It has been an effective 'non-spatial' way of keeping the male and female worlds distanced from each other.

Language creates social reality, not only in concrete terms, but in the abstract. Charlotte Perkins Gilman wrote in 1911 that 'we have not in our minds the concept, much less the word, for an over-masculinized influence'.[24] Thus, before the word 'sexism' was created, there was no clear conceptualization of sexist behaviour and ideology—no *word* for it. Having a word means that little explanation of a phenomena is needed. It also legitimates the concept and creates a bond with others. Cheris Kramarae quotes a panel discussion with Julia Penelope Stanley, who said, 'With language, I can claim aspects of myself that I've denied, express ideas that have been suppressed and tabooed for a long time . . . define my life as real, and I can act to change my life'.[25]

Dale Spender in *Man Made Language* documents the built-in bias and inequality within everyday language, for example the negative connotations of 'spinster' compared to the image of a 'bachelor'.[26] She analyses interaction between the sexes, showing the way men control conversations, interrupting women frequently and dominating the choice of subject matter. Miller and Swift also document studies on the use of sexist language which indicate that from childhood onwards the world is perceived through sexist language which structures women's understanding in an unconscious way.[27]

The issue of the use of 'he', 'man' and 'mankind' to denote all humans is a further interesting case. Though many people continue to see the issue as unimportant, studies like those conducted by Wendy Martyna document the exclusion felt by women when using the generic 'he'. She concludes that apart from the political, consciousness-raising impact of non-sexist or gender-neutral language, its use is vital in enhancing the self-concepts of 'those who have been excluded by sexist language forms and changing the views of those who view such exclusion as mere illusion'.[28] Language needs to include women explicitly.

Interestingly, the so-called generic term 'man' was a rule introduced into scholarship in England in the sixteenth century. In 1553, Thomas Wilson suggested that man should precede woman because it was 'more natural'. In 1646 Joshua Poole insisted that 'man' deserved priority, being the worthier sex. John Kirkby in 1746 extended this to assume that the male gender was 'more comprehensive' and in 1850 an all-male parliament passed the Act which 'decreed that he/man should stand for woman'.[29]

Evidence abounds concerning male control of language. Bernard has indicated its misogynistic base and its hostility to the female world. Within patriarchal language women are designated by terms indicating their debasement, often with an animal association, for example, 'bitch', 'cow', 'witch', 'shrew'. There are 220 sexually demeaning words for a 'promiscuous' woman and 22 words for a similarly sexually active man. Words for parts of women's bodies are continually used as forms of insult: 'what a cunt'.

As well as being misogynistic, the language is a subordinating one where male words indicate power and women's words connote weakness.[30] Thus men 'debate' while women 'gossip'. Words which were introduced to re-label women's position in order to introduce equality into language are trivialized and ridiculed. 'Chairperson', for example, has failed to be a neutral term and is now used mainly for women. Men wittily comment on 'person holes' or renaming a city 'Person-Chester'. Trivializing women's attempts to rename and recreate the meaning of language is indicative of the strength of the threat which those changes intimate.

As well as debasing women, patriarchal language fails to represent many female experiences. There are, for example, no strong words to express the experience of birthing, or the sexuality of mothering itself. Men prefer to see motherhood as pure and precious with no sexual component, and there is no word for the orgasm which sometimes accompanies a drug-free natural birth.

These are important elements of women's oppression and the power of language and literature to control and create society's attitudes and behaviour cannot be overestimated. Catherine Stimpson writes of the power and purpose of literature thus: 'literary texts do have the strength to subvert ordinary modes of consciousness. Naming the strange, unfamiliar, unpalatable, and alarming, they can, potentially, rearrange habitual modes of thought and feeling.'[31]

Roszika Parker and Giselda Pollock have discussed the issue of this ideological control through art. They also discuss language, pointing out that power does not necessarily mean the use of coercion to ensure control. It also operates by denying access to the institutions and practices of dominance. One of these institutions of dominance is language. The important point is that 'we come to know ourselves through being able to use language' which 'embodies symbolically the laws, relations and divisions of a particular culture'.[32] So words help to present and represent the self to the self.

But what effect does it have on women if they use the oppressor's language, as Rich calls it? Mary Daly would argue that it has operated to 'falsify our own self-images'. Woman cannot truly know herself through the misrepresentation of herself she sees in patriarchal language. It is this awareness which has led to the development of feminist dictionaries on the one hand, and a rejection of language and search for the pre-text on the part of some French feminists. Writing of the difficulty of finding authentic female language in which to write the 'truth of women', Bev Roberts says:

Women are linguistically subordinant and subordinated, denied access to power and to participation in the range of public discourse, and experience unique conflicts and ambivalence through the fact that in a male-dominated language environment the female encounters her self and her femaleness in male-defined terms and concepts ... Thus the woman writer finds particular difficulties in self-actualization and self-expression in a language which militates against the assertion of the female.[33]

Trained to use that language which is not strong but suppliant, women have until recently been constrained by having to write, as they speak, in a meek, polite and humble way. They are not permitted, for example, to express anger. If a woman writes in anger it becomes 'bitter'; if she is moving, she displays 'excessive emotion'; and if she speaks against the oppression of patriarchy, she is 'man-hating' or in earlier times, a 'bluestocking'. Her work can be dismissed on all these grounds. And her person is often attacked, as Spender has indicated, trivializing her work as indicative of her 'problem'. So Matthew Arnold could write of Charlotte Brontë in 1853: her 'mind contains nothing but hunger, rebellion and rage'.[34] The woman and her work are rejected on the grounds that these are inappropriate emotions in a good or true woman.

But women's writing since the early 1960s has cast off these assumptions. It strives to shatter the false image and create for the first time the common kernel of women's experience and to communicate it among women. Rowbotham comments on the essential need for this process: 'in order to discover its own indentity as distinct from that of the oppressor it [the oppressed group] has to become visible to itself'.[35]

Bernard points out that until recently, a 'dark "underworld" underlay the female world, a world too awful to speak of, a world that was literally "unspeakable"'.[36] This world contained the reality of the hidden violence of oppression—sexual harassment, rape, incest and wife-beating. These experiences are part of the fabric of women's lives, but have been forbidden the analysis of literature by the rules of the canon. Women no longer regard these rules as valid. As new writing indicates, language can and will be found in which to reclaim and represent the world of the Hags, Spinsters, Witches, and Crones as Mary Daly re-names women. Using old definitions seen in a new perspective and creating woman-centred definitions, Daly reconstructs language. 'Haggard writing is by and for haggard women, those who are intractable, wilful, wanton, unchaste, and especially, those who are reluctant to yield to wooing', she writes, re-energizing the 'obsolete meanings of these words'.[37] Naming and reclaiming language have been major themes in Adrienne Rich's poetry. Time and again she expresses this need with precision, and condenses the power of that new naming:

> I need a language to hear myself with
> to see myself in
> a language like pigment released on the board,
> blood-black, sexual green, reds
> veined with contradictions
> bursting under pressure from the tube.

Currently feminist writers and critics are battling to find ways in which women can regain control and freedom within language. Analysing these approaches, Toril Moi notes that the Anglo-American approach tends to stress the male domination of language that has been discussed above, noting the 'oppressor's language' and seeking to re-name and to place new meaning into existing language. But the French feminist arguments claim that the foreign tongue of male language must be totally discarded in order to find a women's language with which to speak 'woman' into existence.[38]

Hélène Cixous in her essay 'The Laugh of the Medusa' exhorts women to write of themselves: 'Write! and your self-seeking text will know itself better than flesh and blood, rising, insurrectionary dough kneading itself, with sonorous, perfumed ingredients, a lively combination of flying colours, leaves and rivers plunging into the sea we feed'. The re-connection of the language with women will come through the body, which women will reclaim to themselves to love.

Women must write through their bodies, they must invent the impregnable language . . . if woman has always functioned 'within' the discourse of man ... it is time for her to dislocate this 'within', to explode it, turn it round, seize it; to make it hers, containing it, taking it in her own mouth, biting the tongue with her very own teeth to invent for herself a language to get inside of.[39]

Here is vividly expressed the woman writer's need to destroy male control of her through language. It risks however a tying of women to biology, a relationship which would need to be redefined in a woman-centred way.

Writers like Julia Kristeva have sought language *beyond* the text, in the 'pre-text'. This can be represented as the rhythmic babble of onomatopoeic sounds, such as those which occur between mother and child in the pre-language state of the child's development. This occurs before the oppression of the Father's language begins. This semiotic mode, however, is not a woman's language, but 'an eruption in writing that takes the form of breaks within the phonological-syntactic structure of the sentence'.[40] Kristeva sees a necesssity for women to be part of movements which break with tradition and attempt new forms of discourse.

The psychoanalyist Luce Irigaray explores the work of Freud and Lacan. She believes that women need to learn the language of these discourses first, and then to ' "derange" it, bother it, divest it of its power over us'.[41] Again she sees the need for the language to stay close to women's bodies, and works for the creation of a feminine language, attempting to found it outside patriarchy. Hélène Cixous works on writing as a female essence, concentrating on the female body. She wants to move the reader away from the primary concentration on the writer, 'a disabling author-centred empiricism', to the reader, encouraging involvement in text creation. Many feminists find Cixous and Irigaray dangerous in their emphasis on the body and difference, which may lead to a biologism, and decry their utopian aim of a female language. However, their influence has been important in developing a positive concept of female difference. As Jones writes:

Our political pragmatism, empiricism and search for a women's literary history are already intersecting in provocative ways with French interrogations of the politics of discourse, the processes of the unconscious and the outline of a future *á la féminine*.[42]

As language is part of culture, these attempts to reclaim a language for women assume a woman's culture which operates outside malestream society. Bernard has detailed at length the 'female world'. She points out the difficulty of defining culture itself. But, 'whatever evaluation is made of this female culture, it does reflect the inner values, the longings, the dreams of women, and cannot be dismissed as unimportant'.[43] Diaries, autobiography, and women's journals are all part of this cultural experience of women's lives.[44]

Thus women who write do so within the male tradition, but also within a woman's culture. They have a dual tradition in some senses.

This can place them in the position of a divided self: one part constantly defined as 'other' and marginal, while the other struggles to write and is often disguised to avoid detection. Many women writers took on this struggle. But in the last twenty years, the divided self is being reworked too. The self *within* the male tradition is angry and full of self-destructive power; while the woman's culture is re-validating women's power, strength and vulnerability. This can be seen emerging particularly in poetry and will be discussed later in this chapter. Women need to maintain this struggle with language, literature and culture, in order to survive in authentic selfhood. As Kimball writes:

If her language uses the pronoun 'he' as the norm, if her deity is personified as Father and Son, if painting and song lyrics depict women as decorative and sexual objects, if her modes of viewing the world are shaped by male perceptions, a woman cannot form an accurate or positive self-concept.[45]

Silence and exclusion: the material and social conditions of women's writing

Barriers erected by men have made writing difficult for women. The silence of women writers is not indicative of a lack of skill and creativity (actual and potential), but a measure of the material and social difficulties faced by women when trying to nurture that creative spirit. Sheila Rowbotham has written that 'the revolutionary must listen very carefully to the language of silence. This is particularly important for women because we come from such a long silence'.[46]

When Virginia Woolf discussed why there was no Judith Shakespeare, sister to William, she foresaw her fate in terms of an unfortunate love affair and unwanted pregnancy, thus ending her writing career. Woolf often commented herself on the 'strange spaces of silence' which separate periods of women's writing, which she blamed on law and custom. She wrote:

When a woman was liable, as she was in the fifteenth century, to be beaten and flung about the room if she did not marry the man of her parent's choice, the spiritual atmosphere was not favourable to the production of works of art.[47]

The powerless position of women worked to exclude them from basic individual rights, making it often impossible to even aspire to enter 'culture'. For many writers this is still the case. Tillie Olsen documents, with extracts from diaries and journals, the continuity of the experience of disruption and demand placed on literary women. Many women writers did not marry, or married late and did not have children because the time needed for creative work is incompatible with the needs of husband and family as established within patriarchy. Olsen points out that most of the writers she discusses had household help. She quotes Katherine Mansfield: 'the house seems to take up so much time . . . I get frightfully impatient and want to be working . . . well someone's got to wash dishes and get food'.[48] And these, after all, are the jobs of women.

A most productive period of novel-writing for women in Australia was between the wars, and they expressed then the dilemmas of trying to work politically while wanting to write (Jean Devanny and Katharine Susannah Prichard), or trying to service a family or relatives and write. The women were very conscious of the situation of male writers in comparison, with their supportive family environment. Miles Franklin wrote in 1932 that Australian women were 'fast becoming a nation of charwomen. I'm too busy doing chores, myself, to write any more'. As Carole Ferrier comments, 'the usual conditions of women's lives militated against writing'.[49]

The history of women writers is full of such experiences as Sylvia Plath rising in the dark hours before the children woke to write poems; or Colette stooped in the early morning, after satisfying her lover, to light the fire by which to write. The picture is of women snatching at time. Some, like Emily Dickinson, took the way of withdrawal from social life in order to write. Social laws, which women have internalized to a large extent, demand that they look first to the smooth running of the household. Children get primacy, yet are a constant source of distraction and interruption. Neglected partners inspire guilt. Women are trained to place the needs of others first, yet art needs 'wholly surrendered and dedicated lives; time as needed for the work; totality of self'.[50]

Does this mean that women must give up the pleasures of home and family in order to write? Many inevitably answer yes, if that love and family demand the sacrifice of self which denies art. Spacks asks, 'is the cost of achievement the loss of relationship?'. Women's needs are the same as men's, 'for work and love, for independence and dependency, solitude and relationship, to enjoy community and value one's specialness'.[51] These polarities, though, are presented as contradictions for women, who must choose one pole to the exclusion of the other, or face enormous difficulties in pursuing their craft.

Olsen's work in *Silences* is also interesting for its analysis of male writers and their experiences. Constantly they write of the devotion of their wives, who ensure the smooth running of their daily lives, or their inability to write because of the necessity to earn a living—which is a distraction. The work of creation, Olsen concludes, requires constant toil, solitude, a smooth daily life, discipline and self-control. The male writers could mostly devote their full lives to the task, though a couple, like the American poet William Carlos Williams, also had jobs. But these men were of course spared the constancy of household maintenance. And household maintenance does not just mean the *jobs* to be done, but the constant preoccupation of the mind with the *organizing* of the household—as Mansfield says, a mind 'full of the ghosts of saucepans'.[52]

The act of creating literature is difficult and little understood.

Psychologists offer nothing to an understanding of this process, but writers themselves attempt time and again to explore the process. Virginia Woolf wrote in 1929 that 'we are beginning to understand how abnormal is the effort needed to produce a work of art, and what shelter and what support the mind of the artist requires'.[53] There needs to be a 'receptive waiting', a concentration of self and the time to explore the artistic impulse when it comes. This need to work immediately cannot be satisfied if it is dinner-time for the children. And if the moment passes, 'all may vanish as a dream; worse, future creation be endangered, for only the removal and development of material frees the forces for further work'.[54]

Adrienne Rich, American feminist and poet, experienced these difficulties in writing when her three children were young. In discussing the difficulties of poetry she writes:

For a poem to coalesce, for a character or an action to take shape, there has to be an imaginative transformation of reality which is in no way passive. And a certain freedom of the mind is needed—freedom to press on, to enter the currents of your thought like a glider pilot, knowing that your motion can be sustained, that the buoyancy of your attention will not be suddenly snatched away.[55]

Australian poet Gwen Harwood also explores this in 'Burning Sappho':

The clothes are washed, the house is clean.
I find my pen and start to write.
Something like hatred forks between
My child and me. She kicks her good
new well-selected toys with spite
around the room, and whines for food.
Inside my smile a monster grins
and sticks her image through with pins . . . .
Night now, Orion first begins to show.
Day's trivial angers cease.
All is required, until one wins,
at last, this hour. I start to write.
My husband calls me, rich in peace,
to bed. Now deathless verse, good night.
The pulse of song grows faint, and dies.
Out of their pit the furies rise.[56]

That 'freedom of mind' which Rich discusses derives from a confidence in the self. No art can be created without self-knowledge and confidence in the act. Elizabeth Barrett Browning called it 'self-love'. It is what Moers calls a 'miracle of temperament' and it is a miracle in woman when the negative forces against her identity are considered. It requires the antithesis of the self-sacrificing female image; it requires a strong self and ego, an independent mind and a spurning of 'that subserviency of opinion which is generally considered necessary to feminine softness'.[57]

Women's writing in general has been a struggle to develop, and hold strongly to, a positive and active self-definition against the current of compliance demanded of them. Self-redefinition is necessary before a woman can begin to create because she breaks the bonds of culture set by men. Of this process, Gilbert and Gubar write:

For all literary artists, of course, self-definition necessarily precedes self-assertion: the creative 'I AM' cannot be uttered if the 'I' knows not what it is. But for the female artist the essential process of self-definition is complicated by all those patriarchal definitions that intervene between herself and herself.[58]

This strength of self also needs time and encouragement within women, though to the male writer it is his by birthright, purely by virtue of his sex. So Walt Whitman, the American poet, can write a 'Song of Myself' in self-celebration, while Emily Dickinson writes of self-effacement—'I'm nobody'. And poets like Robert Lowell and John Keats can state their desire for greatness after death, which in a woman would be labelled conceit.

Trying to reconcile conflicting demands on their lives, women writers often feel that they failed to achieve well in their writing, fighting a constant battle with low self-esteem. Ferrier quotes Miles Franklin writing to Dymphna Cusack: 'You flatteringly ask how I reconciled the claims of job, personal and literary life. Well, I *didn't*. I have never written except rough fragments which were merely a sample of all that was once dammed-up in me, and which is now atrophying'.[59]

In spite of these impediments to 'voice', women have been writers of some stature in the past. These achievers gained their goals through their talent and skills, but also because of opportunity and tenacity. Exceptional opportunity allowed many women to create an independent self, and some had the fortune to be encouraged by mother, father, lover or friend. Olsen reminds us of the many who had no possibility of success, though the spirit and inspiration was there: poor, overburdened or uneducated women stand little chance of access to the attempt to write. Indeed, as education for women has only been a recent phenomenon in our history, it is a wonder women have been writers at all.

Silence and exclusion: publishing and the critical canon

Men have added to the structured impediments to women's writing through their control of publishing and their dominance of literary criticism which establishes the literary canon, that body of writing which they deem to be the best, most well-crafted, most important and most worthy of posterity. Items in the canon are reprinted constantly and are studied in the education system. Because women's writing lies outside the canon, on the whole, it is not reprinted decade after decade but falls permanently out of print. This means that writers and their work are forgotten and hidden from history. The silence of women in literature is thus institutionalized.

It is essential that women know of their foresisters within literature. Gilbert and Gubar see this as a prerequisite for beginning the struggle of self-redefinition needed by women writers. We need a *'female* precursor who, far from representing a threatening force to be denied or killed, proves by example that a revolt against patriarchal literary authority is possible'.[60] This reclaiming of the past helps to lessen the anxiety experienced by the writer and develops within her a sense of continuity with women writers who went before her: a sub-culture with which to identify. This has been one of the first actions of the recent wave of the women's movement: reclaiming lost women writers. Dale Spender, in *Mothers of the Novel* reclaims the novel form itself as that which women at one point made their domain.[61] Male history has written their contribution out. Feminist presses such as Virago and Pandora Press in England have contributed by reprinting women's writing which was forgotten because it was out of print.

In her book on women's unpublished heritage, *Intruders on the Rights of Men*, Lynne Spender discusses the gatekeepers or guardians of culture. Publishers decide what will be made publicly known and publishers have been men serving the interests of the male perspective. Spender outlines the obstacle race involved in publishing, and the 'heritage of harassment' with which women writers have had to contend.[62] Part of the reason for the exclusion of women has been that they challenged the notion of what a true woman is or should be. Understanding of this reality has led many women, for example, Henry Handel Richardson, Miles Franklin, George Sand, George Eliot, and many more, to assume a male pseudonym in order to be published.

The establishment of women's publishing houses has always been a part of the striving to express female consciousness and experience in women's writing—unacceptable to the male gatekeepers. Early in the twentieth century, women of the Women Writer's Suffrage League in England explored this issue. Elizabeth Robins, the first president, a novelist and actress, argued in 1908 that no woman had been free to write the truth of women's lives: 'to say in print what she thinks is the last thing the woman novelist or journalist is so rash as to attempt . . . Her publishers are not women'.[63] Emily Faithfull's Victoria Press was established in the 1870s and suffrage presses were flourishing at the turn of the century. In 1911 there were 21 regular feminist periodicals in England and a feminist bookshop. Women have constantly fought to be read.[64]

Feminist journals and periodicals have played an important role too. Since the first women's periodical in England, the *Female Spectator* (1744—46), was produced by Eliza Haywood, these journals have continued to appear, disappear and reappear as vehicles for women's writing and women's political ideology. The *Dawn* in Australia was published and edited by Louisa Lawson, a feminist and intellectual in the 1880s and 1890s, and the *Australian Women's Sphere* was owned and edited by feminist and suffragette Vida Goldstein. *Time and Tide* in England was an influential periodical financed and run by Lady Rhondda as a political journal dealing with current issues relevant to women, and it survived for forty years. A radical political journal of its time containing contributions from most of the radical women writers and thinkers of the period, it is 'effectively hidden from enquiring minds' in the inner-suburban depot of the British Library.'[65]

One reason for the continuous cycle of appearance and disappearance of women's publishing and periodicals is that they are the efforts of an individual or a small group of women, and not part of the established economic fabric as male publishing is. Men own the processes of publication and production and only tenacity of effort wrenches part of this monopoly from them periodically. When this occurs, abuse and disparagement is heaped upon the upstarts in order to strangle their opposition.[66]

Decisions made by publishers are reinforced or informed by the opinions of literary critics. These critics draw up the rules for good or bad literature, and work which passes the tests of these rules passes into the literary canon. Writers on the whole do not spare much affection for the critics, who have the reputation of being 'failed' writers, or as Erica Jong describes them 'critics/with their Chipmunk cheeks/full of unwritten poems/better than mine'.[67]

Women who write have faced a critical canon which is male-controlled, so that their work has been judged according to male-defined criteria of literary worth. These criteria demand that there be a certain 'objectivity' and distance of the writer from the work; that the writing be 'universal' in its concerns, by which is meant concerned with wars, battle, death, and so on; and that the writer show an exercise of control over the work.

The kind of censorship of women which has operated in Australia is exemplified in the publication of Barbara Baynton's 'The Chosen Vessel'. Her story was censored by A.G. Stephens of the *Bulletin* in 1896 because it was 'too outspoken' and broke the male tradition of the bush. Stephens changed the title, omitted sentences, and cut the entire third section of the story.

The story itself is highly charged and, as many of Baynton's stories did, it conveyed a hostile image of the bush from the viewpoint of the victim. In the story, the woman is treated as an animal by her husband who leaves her with child alone in their hut. Under threat by a passing bushman, the woman cries for help from a passing stranger. This man, embroiled in his own inner preoccupation with religious virtue, sees her as a vision of Madonna and child and passes by. She is raped and murdered by the bushman.

Iseman hopes that her analysis of 'The Chosen Vessel' may 'help to restore to the vessel a richness and multiplicity of meanings which has been lost through repeated critical attempts at phallic mastery'.[68] Baynton's work demands more attention from critics because of what is still called her 'dissident' writing, and for her exploration of the violence of bush life for women. In terms of the woman writer speaking her exile and her experiences as 'the other', Baynton represents an essential landmark in that alternative tradition women seek. As Lucy Frost writes of Baynton's fiction: 'it is located in a zone of consciousness where terror abides and the self is constantly under threat'.[69]

Within the canon, women are seen as objects of desire or disparagement, or as the muse. It contains male myths about women's bodies, sexuality, and evil versus good impulses. Women have found that they are the 'other' within literature as well as in society. Their own experiences are not represented within what Cheri Register calls the cosmic terminology of literary criticism and male writing. 'The "female experience" is peripheral to the central concern of literature—which is man's struggle with nature, god, fate, himself, and not infrequently, woman'.[70]

The rules of the canon are perpetuated within institutions, such as universities. So students study James Joyce as a 'successful' stream-of-consciousness writer, but not Dorothy Richardson and May Sinclair with whom the form originated, nor Virginia Woolf who perfected it. The canon, written by men, published by men and taught by men defines who is best at representing 'universality'.[71]

Criticism itself is not an objective enterprise. It is the most subjective of crafts and is also an artistic process of a different kind. Through it, the critic actually creates the meaning of the text, recreates and explains the author's intention. The critic then assesses *that* explanation of the text. Kolodny and Ellman have explored the inability of the male critic to understand women's writing. 'There must be', says Ellman, 'two literatures like two public toilets, one for Men and one for Women':[72] one men can comprehend, which is elevated to the canon, and one which they cannot, which is just concerned with 'women's issues'.

The job of the critic in the past has been to interpret, codify, judge and explain a text in order to assess its quality and its rights of advancement to the canon. Criticism has claimed an objectivity for itself which is false, as all critical opinions within literature are based on the likes and dislikes of the reader. Catharine Stimpson writes of the critics' intention to impose 'their will upon a work and to control it. Instead of letting a poem or a novel act upon them, they act upon it'.[73] The history of what Mary Ellman calls 'phallic criticism' has led critics like Diamond and Edwards to suggest that the proper target for feminist anger is not the literature itself 'but the misconceptions of past critics, the received evaluations about literature which , rooted in bias, have for too long passed for disinterested impartiality'.[74]

Women's books have been treated as if they *are* women, with 'an intellectual measuring of busts and hips'.[75] Within this process, a

good book by a woman is one which is free of those traits the critic
ordinarily dislikes in a woman and therefore in her writing. No higher
praise can be found for a woman's book than that it seems *unlike* the
writing of a woman and could be passed off as a man's book. These
critics strain to see femininity within the text, which they then disparage.
If the text deals specifically with women's experience, it is naturally of
dubious quality, not being 'universal' in its subject matter. Its 'femininity', moreover, must be of the right kind—no anger, bitterness,
raised voices or dissatisfaction with the female lot.

The pressure on women from critical schools not to represent their
true lives has meant the creation of a hidden text, only now being
excavated by feminist critics, or a literature revealing a confusion of
identity and anxiety. When feminist critics read this work, a counter-
tradition of women's writing begins to emerge. For example, Ellen
Moers wrote *Literary Women*, detailing a female tradition within women's
work based on specific female experiences, such as childbirth and
economic dependence, everything in fact 'special to a woman's life'.[76]

Patricia Spacks developed her theory of the 'female imagination'
within this tradition, claiming its specificity in contrast to the male
imagination. Though many debate this point, her work is interesting
particularly for its reconstruction or revision of woman's so-called
masochism, narcissism and passivity within the novel, denoting them
to be 'strategems for maintaining personality' in a situation where
direct tactics were impossible. When discussing, for example, the
masochism of Maggie in *Mill on the Floss*, Spacks writes of such
heroines: 'their capacity to accept, even welcome, unhappiness derives
from their refusal to compromise, their unwillingness to conform to
social definitions of what should constitute happiness, their determination to preserve the integrity of self'.[77] She comments that everywhere
in women's writing runs their anger about their condition, which
Spacks (unlike Virginia Woolf, who was irritated by it) finds a source
of women's creative energy.

Elaine Showalter wrote *A Literature of Their Own* about British
women novelists from the Brontës to Doris Lessing and developed a
theoretical framework which was to be used frequently over the
following years.[78] Uncomfortable with the notion of female 'imagination'
which she felt was close to stereotyping women writers, Showalter
maintained that there was a tradition of women writers which could
be discerned as developing through phases, related to the social conditions in which they wrote. The *Feminine* stage (roughly 1840 to 1880)
she described as a phase of imitation of the dominant tradition with
an internalization of the standards of the canon and of women's
'appropriate' behaviour as writers. The second phase is *Feminist*
(1880–1920)—a protest against these standards, advocacy of minority
rights and the determination to claim autonomy. Showalter notes that
protest fiction written in support of any other oppressed group, such
as child labourers, working men or slaves, was permittable in women,

and in fact channelled much of their anger and frustration into a more acceptable form than writing about women's oppression would have allowed. The third stage she labelled *Female* (1920 onwards) and stresses the search for identity and self-discovery through a turning inwards, in contrast to the oppositional literature of the feminist phase.

These forms of analysis created a renewed perception of woman as a cultural group. Without this perception, critics could 'miss or misinterpret the themes and structures of women's literature, fail to make the necessary connections within a tradition'.[79] One of the revelations available with this new perception is the understanding of the pressure on women writers to hide their themes and shroud their statements within the main theme of the novel. Again, Elizabeth Robins had written in 1908 that though she had 'access to a rich and as yet unrifled storehouse', the woman writer had

cogent reasons for concealing her knowledge. With that wariness of ages which has come to be instinct, she contented herself with echoing the old fables, presenting to a man-governed world puppets as nearly as possible like those that had from the beginning found such favour in men's sight.[80]

Feminist critics began to turn their ear to the sound behind the echo.

One example of this is Gilbert and Gubar's analysis of 'the madwoman in the attic' and her role in the tradition. Here, the madwoman is seen as the double of the 'heroine' who can speak the unspeakable, reflecting these 'uniquely female feelings of fragmentation' and a sense of the contradictions between what women are and what they are supposed to be.[81] So, for example, Charlotte Perkins Gilman's *The Yellow Wallpaper* (1899) depicts brilliantly from the 'inside', as it were, the recovery and the assertion of self using the vehicle of madness in the main character, with whom the reader identifies. So women used pseudonyms, the hidden plot, and traditional plots which hold within them the story of women's suppressed rebellion.

In the twentieth century, women's novels have burgeoned, taking on a new power and resonance in the search for identity. Science fiction and utopian/dystopian novels have been used by women to project their loathing of patriarchal violence and aggression, and their yearning for a woman-centred, loving and creative society. Many of them stress the power relationships between the sexes and carry a sense of urgency. The historical form too is rewritten in order to represent the saga of a woman's tradition in realistic terms.[82] As Stimpson has pointed out, women's texts deal primarily now with the world of women: 'In some texts, a world of men and a world of women impinge upon each other—with love, in terror, with suspicion. In other texts women have left men, for a while, or forever'.[83] Stimpson also discusses the lesbian novel in the English tradition, whose authors face a double burden in writing both within a patriarchal culture and within a culture of heterosexuality.

Australian women writers too have suffered efforts to extinguish their presence from the history of Australian writing. But books such as Drusilla Modjeska's *Exiles at Home* have provided a wealth of information on women writers between the wars, their relationship to socialism, feminism, and the peace movement. Half the novels published between 1928 and 1939 were by women. In 1928, the *Bulletin* offered a literary prize of £1700 which attracted 540 entries. First prize was shared by Katharine Susannah Prichard and M. Barnard Eldershaw. Modjeska quotes Geoffrey Serle (1973) on that period: 'however unconventional one's taste or ranking, it would be almost impossible to deny that most of the best novelists are women'.[84] It is interesting that the novelist M. Barnard Eldershaw was actually a literary liaison between two women, Flora Eldershaw and Marjorie Barnard. The two wrote many articles and stories together in successful collaboration.

One of the most important and neglected writers was the literary critic Nettie Palmer. She created a network by correspondence, of mostly women writers. She encouraged and challenged them, offering critical advice, and established a prolific communication on literary and social issues. Nettie Palmer was distinguished in her career, but most Australians are more familiar with her novelist husband, Vance Palmer, who in the main she supported in order that he might write. To do this, Nettie gave up her own literary ambitions to stay with criticism, which was more remunerative.

For many Australian women writers, the only escape route from their double alienation seemed to be exile. For Miles Franklin, the exile from Australia was marked by overwork and self-doubt. Her years in America and England have tragic tones: her political commitments increased as did her doubt about her literary creativity.[85] But she was active and effective in the women's movement and it is frustrating that these experiences were not recreated within her literature.

Henry Handel Richardson also lived in exile, as did Christina Stead. This was not an easy thing for women to do. They must first save the money to travel. Stead, in her mainly autobiographical *For Love Alone*, chronicles the painful and difficult story of this particular escape. It is the story of Theresa's quest for herself, for her creativity and for self-chosen love.

The title of Modjeska's book is telling. For those who remained in Australia, the experience was one of being an 'exile at home', alienated by the patriarchal culture of Australia, and made invisible within the mainstream of Australian writing. Feminist critics and writers are, however, beginning to struggle with the male stranglehold on Australian literature and to reclaim, reassess and rewrite the lost tradition of women writers which is necessary for those who currently write.

Women poets in Australia, with the rare exception of established 'names', are continuously and systematically excluded from the canon. A consideration of anthologies of Australian poetry attests to this fact. Even recent anthologies continue this tradition: in John Tranter's *The*

New Australian Poetry of 'twenty-four poets from Australian poetry's most exciting decade' two women appear.[86] This poetry comes from the closing years of the 1960s and the poets have been labelled 'the generation of '68'. It is an interesting selection of work by male poets. Considering the few women represented (two out of twenty-two), it is not surprising that in his survey of the major influences of the period, Tranter neglects to mention the women's liberation movement at all. No indication of this massive stimulation to the consciousness of women is made, and the reader could assume that Australian women were not involved in it. This lack of women's representation is carried through various anthologies: for example, those of Robert Gray (Angus & Robertson) and Les Murray (Oxford University Press).[87]

Editors have tremendous power in controlling the direction of poetry. Rodney Hall points out that the editors of magazines and journals are 'a poet's most formative audience'. By the very act of accepting or rejecting a poem, they markedly shape the author's next submissions, not to mention what advice they might offer on revising manuscripts'.[88]

In his autobiographical *Cutting Green Hay*, poet Vincent Buckley details his experiences with the intellectual and political movements associated with Melbourne University between 1945 and 1965. Women are absent from the book. Judith Brett has pointed out that there are some 200 men in the index and 17 women, 12 of whom receive only a single mention in the book. Brett poses the question, where were all the women and what role did they play in ' "the movements and cultural conflicts of Australia's great decades?" From the evidence of this book—none'. When women do appear they are relegated to the private domain of the family. Brett writes 'I found the ease with which Buckley sets his memoir in the "public" world one of the most unsatisfactory aspects of the book'.[89] Vincent Buckley was the editor of the *Bulletin* poetry pages which he took over from Douglas Stewart in 1961. He transformed Australian poetry through his publication of new voices; but it could be asked how did the male-dominated view of the world expressed in his autobiography affect his selection of poems to be published in the *Bulletin*?

Poetry as resistance

Though women struggled and made inroads into the field of the novel, poetry has resisted more strongly the inclusion of women. There is some reclamation of women poets, but there were not as many successful poets as novelists. In Australia, Lesbia Harford and Marie Pitt, writing mainly in the years 1909 to 1927, are two poets reclaimed from history.[90] Again, they were writing between the wars, a flowering period for Australian women writers. Cuthbert indicates this was partly the result of the forty years of a growing feminist awareness and an increase in the liberty of women. Pitt, as did many poets of the time, idealized the bush and pastoral life, yet she was aware of the unequal position of women in general. She explored in her poems about women the theory that man's first act in the world was the enslavement of women. She also sees an enmity between two classes of women: ruling- and working-class. Pitt's poetry indicates that she saw the power of women as lying in their ability to reproduce and in their moral superiority.

Harford, on the other hand, did not idealize rural life. With a degree in law from Melbourne University, she chose to work in factories and joined the Wobblies, the International Workers of the World, a radical labour movement. She wrote more in line with the socialist feminists of today and the poetry was personal, confessional and of a metaphysical nature, though regular in form. Cuthbert comments that 'the tone and structure is somewhat reminiscent of the poetry of Emily Dickinson'.[91] Some of her work is quite radical, and deals with the emotional, psychological and biological aspects of being a woman; factory life and the experiences of work for women; and the contradictions between the desire for independence and the need for love.[92]

But poetry has been a difficult craft for women because of the specifics of poetry and the constraints on women writers discussed above. Where the novel could be created over a long period and could be more easily disrupted and returned to, the moment of poetic creation demands an intensity which disallows interruption. Women's lives do not often allow time and silence for such concentration.

Poetic creation involves a waiting, listening period, followed by intense creation. Women have been seen as incapable of this for two reasons: one is self-discipline, which women are perceived not to have, and the other is the 'excessive emotionalism' of their work which Bev Roberts attributes to an 'underlying anger at the male appropriation of poetry'.[93]

Since the early 1960s the shape and substance of women's poetry has changed. The first poetry emerging from the recent wave of the women's movement dealt primarily with the search for self-identity and used as its substance the personal lives and experiences of women. It was characterized by a direct, accessible form and a personal voice. The language was literal rather than symbolic. Though that is still true of much current poetry, the newer works have wider symbolism within new meanings. For example, woman and nature are reconnected but with the intention not of sentimentalizing that connection, or tying women to biology, but of creating a source of strength and wisdom for women. The imagination, dreaming, and the creation of myth are primary concerns. From an explanation of social and personal relationships, women poets are moving into 'the space of mind' and the unconscious. This development of social myth Janeway sees as a primary direction.[94]

The search for identity has been a primary concern of recent poets.[95] Roberts points out that whereas the novel form allowed women to come to terms with the social, 'in poetry they come to terms with the self, not divorced from the social, but with the potential of transcending it'.[96] So the American poet Adrienne Rich writes of her feeling of sisterhood, separate yet coming together with other women; its positive affirmation reaches beyond the individual's experience alone —

> yet never have we been closer to the truth
> of the lies we were living, listen to me:
> the faithfulness I can imagine would be a weed
> flowering in tar, a blue energy piercing
> the massed atoms of a bedrock of disbelief.[97]

Rich herself has commented that the drive for self-knowledge in the woman poet is more than a search for identity, 'it is part of our refusal of the self-destructiveness of male-dominated society'.[98]

In exploring the question of identity in American women's poetry, Gardiner points out that 'female identity is a process' and that for each aspect of identity which men define, 'female experience varies from the male model'.[99] Women, for example, construct a more direct relationship with the reader as opposed to a *delivery* of the text from the male poet. There is a sense of urgency and excitement in this process.

The autobiographical nature of women's writing has enabled critics to dismiss it as 'just about their lives'. This does not happen with a male confessional poet, for example the American Robert Lowell. Lowell's poetry is directly autobiographical. Yet his work is assumed to be transcending the personal, and to be representative of his time. It is, in fact, representing the experiences of many *men* of his age. Rosenthal notes that the personal crisis of the male confessional poet is felt at the same time to be 'a symbolic embodiment of national and cultural crisis'. The male confessional poet then, explores his own psyche, but 'observes himself as a representative specimen with a sort of scientific exactitude'.[100] In contrast, women poets aim for a direct hit on the reader in personal/emotional terms.

Critics continue to do little to educate themselves about the tradition of women's poetry, the source of which differs from that of the male poet. They continue to deny the validity of women's experience as equally representative of the 'world' as man's. In Australia this is particularly the case. John Millett's impressive collection of poems *Tail Arse Charlie* is a sensitive and successful representation of the war experiences of a man (the author) in the air force. They are then, representative only of those experiences and they are directly related to the persona, 'I'. Yet a review by Kerryn Goldsworthy comments thus: 'Perhaps the effectiveness of the poems derives partly from the fact that Millett has no particular wheel-barrow to push'.[101] This implies an objectivity in the stance of the poet, which is naive and misleading. In contrast, a woman reviewer of my own book *Filigree in Blood* comments that the poet 'displays quite a wide range of concerns, though clearly she should be seen mainly as a woman poet much concerned with women's issues'. After commenting favourably on some of the poems, the critic concludes 'the best are those in which the poet *forgets both herself and the harrowing problems of our time*, and dramatises or creates situations in which *other* authors have found themselves'(emphasis added).[102]

Both women reviewers collude in the rules of the canon, by drawing back from the personal in the poetry. The male recitation of his experiences is classified as objective and universal; but a woman's recitation of her experiences of the world is subjective and limited to 'women's issues'. Further reviewers of *Filigree in Blood* noted that the poet 'confronts life and her statements are frank and personal, at times raw and always with that degree of integrity which makes one pay attention'[103] and that the work had 'emotional authority'.[104] It is this 'rawness' and 'integrity' in women's experience which is not 'decent' within the canon.

In terms of the self, the two sexes write from a different position. The male is assured, as we have seen with Walt Whitman. He *knows* who he is. But for the woman poet, self-definition is the goal, not the starting point. She aims to discover the lost, fragmented and broken self, and within it, a certainty and a tradition. In evidence of this, Gilbert quotes a number of self-defining statements which recur in a cross-section of female poets, as if the poet is trying on these definitions to measure the fit. This is again represented in the search for a 'name': 'if they ask me my identity/what can I say but/I am the androgyne/I am the living mind you fail to describe/in your dead language'.[105]

A difficulty in the poet's search for identity is the sense of having a split self or two selves: one by which the world knows her, and the other, her secret name, in which are hidden 'her rebellious longings, her rage against imposed definitions, her creative passions, her anxiety, and—yes—her art'.[106] This second self can be her strength or can lead her into self-destruction.

The poetry of self is thus very personal but communicative, a parable of women's experience. Janeway comments that it cannot be judged critically without understanding that 'part of its impact comes from its intimacy, and that this impact derives also from a shared moment in time'.[107]

In her discussion of confessional women poets in North America, Rodgers points out the range of their experience and emotion which is displayed, and the breadth of their understanding of women's experience. She also highlights some of the major themes which recur in this poetry, for example anger, self-destruction, madness, frustration and the failure of love to meet women's needs.[108] As indicated above, these feelings have been disallowed to women writers, but the new poetry demanded their release. Adrienne Rich has chronicled women's anger well: anger at exclusion from the world, at being controlled and oppressed, at male violence. In 1972 she wrote of:

> my visionary anger cleansing my sight
> and the detailed perceptions of mercy
> flowering from that anger.[109]

Yet the anger is incorporated as part of the whole self. With it endures a faith and a compassion. Still writing of it in 1978, Rich wrote:

> Anger and tenderness: my selves.
> And now I can believe they breathe in me
> as angels, not polarities.
> Anger and tenderness: the spider's genius
> to spin and weave in the same action
> from her own body, anywhere—
> even from a broken web.[110]

And so Rich continues to 'dream of the common language' which will bring community and caring: 'we know now we have always been in danger/down in our separateness'. She herself has said of Sylvia Plath and Diane Wakowski's work that it is 'the woman's sense of herself—embattled, possessed—that gives the poetry its dynamic charge, its rhythms of struggle, need, will, and female energy', which were previously unavailable to the poet.[111] Within that struggle, poets grapple with their own reticence at exposing their 'self' and at the risks involved. The concept of woman giving birth to herself has echoed through the poetry. Denise Levertov wrote: 'The poet is in labor. She has been told that it will not hurt but it has hurt so much that pain and struggle seem, just now, the only reality'.[112]

The pain of women's experience has gradually been chronicled by the poets. Jessie Bernard notes the shock with which the literary critics received these blasts from women writers, as female culture was 'revealed as a cloak for pain'.[113] Within this poetry, madness and self-destruction were unleashed. If man has defined woman, I will destroy her, cries the poet. Breakdown has often been experienced by women writers, though madness itself is rare, as is suicide, contrary to the theory that women writers have to kill themselves in the end.[114] But the constant imposition of a male-defined identity and the dislocation of the true self, creates an internal tension and friction difficult to bear.

The madwoman is useful again in poetry for expressing the 'truth' of experience. She can be excused for her inappropriate behaviour. Like anger, the suggestion of madness is powerful, instilling fear into the instigator of the emotion. Plath's poetry is strong in its anger and hate, and its vision of the sickness of humanity with little to redeem it, has even greater impact due to her self-inflicted death. Anne Sexton's work was more a mix of conflict and celebration, though the despair was apparent: 'I was tired of being a woman,/tired of the spoons and pots,/tired of my mouth and my breasts,/tired of the cosmetics and the silks'.[115] Searching for a continuing balance within herself and for some peace and reconciliation of her own conflicts, Sexton touches the hurt places of women's unspoken pain. But for her there was a cost too, and her last poems were 'written from the extreme knife-edge of self-slaughter'.[116]

For Plath and Sexton and for others, the experience of madness became a 'metaphor for the absurdity of the rest of life'. It was 'a limit toward which victimization and powerlessness push women'.[117] But it also represented a freedom and its coming could be celebrated as a revolt. It could be seen as a taking back of control into women's hands, an act of self-assertion.

This period in women's poetry had to end. The scenarios of madness, death or compromise had expressed women's rage and powerlessness effectively. A movement toward recreating the self began. One of the methods used in this process is myth-making, where myth is reimagined female. Much of this poetry is related to the second burst of women's writing since the 1960s which delved further into the unconscious and the imagination.

Patriarchal myths are a strong and powerful tool in the representation of power relationships between the sexes. They are a kind of history: 'stories that a culture tells about itself to explain its meaning; and while the situations may be imaginary, the truths are real'.[118] They belong to 'high' culture and speak of dreams, forbidden desire, the psyche and the unconscious. Ostriker points out that women do not share the modernist nostalgia for a golden past culture and their myth-making is rooted in female self-exploration. So while the initial burst of writing in recent years was in reaction to male culture and poetry, this second rush is activating the female unconscious and the female past.

Ostriker calls the process 'revisionist myth making' which means that a tale or myth is altered and appropriated by women in order to tell the story from a woman-centred perspective. So Canadian Margaret Atwood writes the story of Circe and Odysseus from Circe's mouth in *Circe/Mud Poems*. To Circe, whose life is circular and cyclical, Odysseus' purpose seems linear, violent, and she asks:

Aren't you tired of killing
those whose deaths have been predicted
and who are therefore dead already?
Aren't you tired of wanting to live forever?
Aren't you tired of saying Onward?[119]

Mona Van Duyn recreates Yeats' 'Leda and the Swan' poem with 'Leda' and 'Leda Reconsidered'. In the latter, Leda is an active force whose presence influences the mythic relation:

She had a little time to think
as he stepped out of the water . . . .
her hand moved into the dense plumes
on his breast to touch
the utter stranger.[120]

This is in comparison to Yeats' picture of the ravished and passive
Leda the victim:

How can those terrified vague fingers push
The feathered glory from her loosening thighs?

Women poets turn to the great myths and retell them; they personify
the goddess as all positive female. The strong, aggressive female can
be identified this way as 'truth', rather than as 'deception' as in male
myths.

The issue of the muse often arises in this context. 'Woman' has
been identified as the muse for male poets. But is she the muse for
women poets? Can man be the muse? What qualities could he possibly
represent that would initiate a creative source? Women have turned to
each other and to themselves. Levertov even creates a muse from a
sow, which is the moon and is herself, with whom 'in the black of
desire/we rock and grunt/grunt and/shine'.[121]

With their breaking into myth, women poets have returned to a
stronger symbolic mode. Symbols which were previously male-defined
are recreated female. Montefiori and Kaplan discuss flower imagery
in this sense.[122] Symbolic images of the body have also been reworked.
Those traditionally associated with women's bodies are often retained
with their gender identification, but are transformed so that 'flower
means force instead of frailty, water means safety instead of death,
and earth means creative imagination instead of passive generative-
ness'.[123]

When men write of women's bodies they do so in sexual terms and
in relation to men. So the *acts* of the body are emphasized, rather than
its relationship to the self which resides within. To quote Elizabeth
Hardwick, 'if you remove the physical and sexual experiences many
men have made literature out of, you have carved away a great hunk
of masterpieces'.[124]

To a large extent, in fighting the negative male definitions of
woman's body, women's writing has striven to immerse itself in
anatomy—to write from inside the female body. This immersion in
bodily femaleness has been particularly noticeable in the work of
women poets. Ostriker analyses the body language within poetry,
finding that in the last two decades particularly, women poets have
used anatomical imagery more frequently 'and more intimately than
their male counterparts'.[125] In fact, this more intimate exploration
some critics find obsessive and distasteful but it is intended to tear
away the maudlin sentimental wrappings into which male poets have
placed women's bodies. It is also intended to challenge the stereotypes
of femininity.

So women poets rewrite the flesh and the spirit which man has set asunder. Ostriker considers three attitudes of women poets to the use of body imagery: rejection, ambivalence, and affirmation. These deal with the suffering inflicted on women's bodies by male violence or by disease; the ambivalence of a body which can create life but at a great cost to both body and soul; the physical self as a place to begin reuniting private and public issues; and yet, the beauty and strength of the flesh, and its continuing connection with all women.

This brief coverage only touches the breadth and depth of women's poetry at present. It neglects other important concerns such as love. Love between women and men is seen in its limitations, love between mother and children explored in its difficulties, the love between women recreated as friendship or lovingly sensual. The complexity of women's poetry also remains uncovered here; poetry such as that by Marge Piercy, which contains beauty, the sensual, strength, anger, passion and brilliant craft. Her poem 'The Right to Life' is an example of the meeting of politics and the personal, in a well-crafted unforgettable statement.[126]

Nor is there space to investigate 'dreaming' as it is used in poetry, which perhaps serves a similar purpose to that of science fiction within the novel form. Juhasz has commented that the newer poetry is more difficult than the earlier accessible work and the reader needs to work harder, but 'so did the poet work harder to find words and forms that could speak of the non verbal, transcendent, and profound layers of experience that she has gone on to unearth'.[127] She foresees the beginning of the third wave of women's poetry, in which new images and a new language will be developed. Cixous too has faith in poetry as part of woman's new self-identity. She writes:

But only the poets—not the novelists, allies of representationalism. Because poetry involves gaining strength through the unconscious and because the unconscious, that other limitless country, is the place where the repressed manage to survive.[128]

The self is essentially a part of the creation of art and cultural representation. By denying women autonomous selfhood men ensure their absence in the creation of culture. Women continue to resist that exclusion.

4

The Labours of Women:
Self-definition through Work

Identity and work

The work people do helps to define their value in society, and their sense of identity. In our society, paid labour is valued over unpaid work, which tends to remain invisible and is primarily done by women. There is also a sexual division of labour within the paid and unpaid workforces which contributes to the definition of woman as different to, and of less value than, man. Women are defined by occupation as a serving and servicing class, even when paid. Because of structured inequalities with respect to work, women as a social group do not have economic power and independence: economic dependency forms part of the patriarchal definition of woman.

Exclusion of women in the naming and definition of 'work'

Women constitute half the world's population; are one-third of the world's formal workforce; do two-thirds of its work hours; carry out four-fifths of all informal work; receive one-tenth of the world's income and own less than one-hundredth of the world's property.[1] Yet some women respond 'No' to the question 'Do you work?' Why? Because the very nature of work itself has been defined by men in their own terms in order to value *their* labour. Men work, and are productive, while women are 'just' housewives or 'just' secretaries, clerical workers, mothers and the emotional support systems of patriarchal economies. Women never retire because they never really work, according to men.

There has never been a woman who did not work. But there have been differences in the kind of work they do and whether it is paid for or socially valued. Trapped within a love and duty ethos (see Chapter 5), women have been systematically excluded from the spheres of the economy which would give them real power to negotiate the terms of their existence. Without a real part in the control of exchange, women as a group are reliant on the 'generosity' of men as a group and often on the individual man. Considered an act of love or duty, women's work in the home is seen as non-productive and therefore worthy of no remuneration.[2]

Economics, as Leghorn and Parker point out, is 'simply a system of distribution of the work, resources, and wealth of the society', i.e. a way of deciding who will do the work and who will benefit from it.[3] The definition of economics and work is male-biased. Novarra comments that 'J. K. Galbraith has pointed out that women can go through a college course in economics without once realising that members of their sex have a vital, unacknowledged and unenviable place in the production–consumption cycle'.[4]

The characteristics of male-defined work revolve around production. Work is only such if it is 'productive', if effort is expended toward a purpose, and if it is recognized by payment. Thus, for example, voluntary philanthropic work which is constantly done by women is seen as non-productive and therefore not defined as work.[5] If it is not work then it need not be paid for. Leghorn and Parker discuss women's agricultural labour, the value of which was gradually eroded by colonization in many countries. In sub-Saharan Africa, South-east Asia and parts of Latin America, women are 50 to 90 per cent of the agricultural labour force. But even when they are producing most of the food for their families, their work is not considered 'productive' because the exchange takes place within the family.[6]

Characteristics of male-defined work include the pursuit of advancement; an unbroken record of work; hierarchical structures; self-interest in the workplace (men rarely work for 'nothing'); and an environment which coalesces with the masculine values of the male stereotype, creating a 'harsh, competitive, uncaring' work situation.[7] The value system of the male economy includes profit at the cost of others; product as all-important; money as the measure of worth; exploitation as cleverness; competition as healthy; and acquisition as desirable. The concept of 'success' discourages workers from valuing the work itself. Novarra points out that advancement in career terms mostly leads men away from the characteristics of the job which were initially its attraction.[8]

The gain to men from this work definition is power through access to capital, a sense of control of personal resources, and a self-definition reinforced by the social networks of the workplace. Paid work also enables men to exercise power within their individual households, devaluing the 'life-work' which women do.

But what work *do* women do? Women's labour is complex in that it has primary productive uses in the market place, in the household, and in child-bearing. Thus, the strong division between home and workplace which men experience, and which they wish to impose on women, is not so clearly delineated for women. The boundaries, as Matthews has commented, are permeable. Game and Pringle found this with, for example, nurses, who are made to feel guilty if they take time off to look after their sick child when they should be looking after other sick people. Women then try to differentiate between the two power structures of work and home, hoping to achieve more power at home than at work.[9]

Matthews clarified two dimensions to women's work which differ from men's, and which have been neglected in past approaches to women's labour. The meaning of women's work is more complex than men's in the dimension of time. Thus age, marital status, the size and age of a family, 'as well as class, skill and occupation must be introduced into any notion of work before it becomes meaningful as an analytic category for women'. In addition, the dimension of 'space' is crucial. Judging the performance of women on masculine standards, for example in statistical participation rates, wage rates, promotion rates, has found women 'wanting'. It looks for women's participation in men's space. It leads, for example, to the invisibility of the paid domestic labour of women in the nineteenth century, which involved a quarter of the female population of France, England and Australia. Looking for women in the 'wrong place' has allowed the idea to emerge that women *entered* the workforce in the 1950s. But they simply became more visible by doing work similar to men in their spaces: the 'phenomenon of married women earning money, i.e., working, was not new'.[10]

Men have not acknowledged the involvement of married women in 'an informal economy', where they do paid work *compatible* with their domestic labour. Matthews gives five examples of this paid work in the pre-1950s in Australia: the private sale of skills, for example laundry and piecework; the private sale of home produce/commodities, for example eggs, jam, lace and wool which were surplus to household needs; the provision of lodging, bed and breakfast; the operation of small businesses in or near home, usually a family mixed business, millinery, or teaching music; and 'outwork', that is, the piecework most common in textiles, which was most closely aligned with the mainstream economy.

These areas of women's work existed through the nineteenth and twentieth centuries and were reduced during World War II, when the absence of men at war opened up mainstream-economy positions for women. After the war women experienced pressure to return to the home, and many did. In addition, the 'informal economy' was intruded upon by capitalist mainstream economy. Each of the five areas named above expanded into the capitalist market-place, with the exception of bed-and-breakfast work, which faded out. Thus women still did essentially the same work, but because it had become mainstream there appeared to be a sudden influx of women into the economy.

Throughout the literature on women's work a recurring theme is the relationship of women's domestic labour and child-rearing to her market labour. Novarra sees a constancy in women's labour which pre-dates the money economy and came from necessity. She names six tasks which women traditionally have done and which 'cannot be left undone if the human race is to survive and life is to be tolerable'.[11] The six tasks are: reproduction; feeding children and maintaining home life; clothing people; tending the weak and frail; the education and nurture of the young; and creating the comfort of a home base. These tasks are now for the most part institutionalized, but women still do them; for example jobs like catering, nursing, education, cleaning and textile work are primarily women's jobs. Paid work mirrors women's unpaid work, and is primarily the kind of work which is repetitive, or servicing and of low status.

A frequent characteristic of women's work is the expectation of sacrifice which is tied again to love and duty. For example, women have traditionally eaten less food or less nutritional food when times were hard in order to feed their families. A full meal is expected by a man because he 'works'. 'Making ends meet' has been women's job world-wide. To do this they cut corners on their own needs or expend large amounts of time stretching resources by, for example, doing comparison shopping. This work is invisible to men, who expect it and benefit from it. A survey in Britain in 1975 showed that although prices, and wages, had risen by 26 per cent, only 25 per cent of husbands surveyed had given their wives an increase in housekeeping money and the wives were still expected to 'make ends meet'.[12]

Individually and institutionally, men control women's behaviour to ensure a maintenance of optimum labour. Often financially dependent due to their young children, excluded from waged labour unless 'allowed' to work or forced to through necessity, controlled by patriarchal structures which define 'fit work for women', influenced by the ideology of love and duty, women's twenty-four-hour-a-day labour services men. Domestic labour is constant, with no pay, no vacations and no assured permanency of position. Little is done to improve this situation, and 'men have structured their economics on the foundation of women's slavery' because men benefit from it.[13] The catch-cry of waged women that 'I need a wife' expresses the servant status of that position clearly.

Women's unrecognized domestic labour is a major area of exploitation. But in relationships with men, women's *emotional* work is even further exploited. The social expectations of women in love involve sacrifice of self and one's own needs. In heterosexual relationships men usually rely on 'a mixture of love and exploitation'. The ideal of love for women convinces them that 'nurturing work should be done for free'.[14] This ideology makes it difficult to claim payment for work in the home. Such a claim is made to seem mercenary and coldhearted. A woman's unpaid work does not only involve cleaning, cooking, washing and ironing alone, but also organizing the functions of a family unit to ensure that the needs of all members are met in the best possible way, often with the exception of herself. Women construct an emotional support system for husbands, children and elderly parents, directing their emotional lives, supporting and consoling them. Men live their emotional lives vicariously through their wives. And this enforced martyrdom of mothers, which may become self-destructive bitterness, is not rewarded by family or society. The personal is certainly economic.

In her role as 'emotional shock absorber and confidant' woman is required to give constantly. Novarra comments that 'an increasing number of men desire their wives to be receivers of the psychological waste products of the day's work'.[15] And this labour is also expected of women in the workplace: the secretary, receptionist, saleswoman, bank teller. And if they stop being eternal 'soothers', women find, as did the New York secretaries and receptionists who launched a massive 'smile strike', that they are continuously asked: 'Well, what's wrong with you?' in a tone which says, 'Why aren't you making me feel better?'. As Game and Pringle comment, 'In some ways the emotional and sexual bonds seem harder to break away from than relations of economic dependence and exploitation'.[16]

Along with the control of their bodies, economic control of their lives is the most crucial goal for women if they wish to be freed for independent living. Leghorn and Parker argue that all social institutions have an economic basis: for example, access to education for women, reproductive rights, concepts of prestige and status, motherhood and beauty, are linked economically. Access to paid labour and therefore power also constructs access to these other social institutions. But women are systematically excluded from such power. In all societies, women's work, invisible to men,

sustains the economy and subsidises the profits, leisure time and higher standard of living enjoyed by individual men, private corporations and male-dominated governments ... without the corresponding economic, political and social control over their lives that such a crucial role should mean.[17]

The development of 'men's' and 'women's' work
Public and private spheres
Women are identified with domestic labour, appropriate to their 'supportive' role in the family. How does this come about? Cross-cultural studies of less industrialized countries have been one way of attempting to estimate whether home labour has always been exclusively female. In most cases, because women are closest to child-bearing and -rearing, they are attached more firmly to home base. This lack of mobility makes it more difficult to be involved in bartering if distance is involved, or in employment.

However, the power given to women's labour differs within societies, depending partly upon how closely the private and public spheres are connected. An argument can be developed that the greater the separation between these two spheres, the easier it becomes for men in the public sphere to name their work as more valuable; to control women by allowing/disallowing their access to necessary resources; to devalue women's home-based labour as 'unproductive'; and therefore to oppress women.

The analysis that the increasing division between male/female spheres decreases women's power has focused attention on the impact of industrialization and capitalism. The separation of the work world from home 'has been identified as one of the most significant characteristics of industrialized society'.[18] Before the industrial revolution, the family formed an economic unit of production. Husband, wife and children worked together to support themselves. Labour was intended for the survival of the family unit, not necessarily to create surplus and therefore additional capital. Within this unit, women did essential work: usually involved with feeding (for example, the dairy work), clothing (for example, spinning and weaving) and the creation of household essentials (for example, making candles and soap). Control of the household was still within male hands, but women's work was respected as crucial to the household and no man would marry other than a good worker.

In their detailed study of the pattern of women's work from the seventeenth century onwards in England and France, Tilly and Scott name the pre-industrialization unit as the 'family economy'. They write that: 'high fertility, high mortality, a small-scale household organization of production, and limited resources meant that women's time was spent primarily in productive activity'.[19]

During industrialization the size of the productive units grew and increasingly people worked for wages. Women were still involved in domestic labour with added episodic wage labour (children also worked outside the home). This period introduced the 'family wage economy'. It also saw the development of an enlarged middle-class introducing an ideology of domesticity, where the status of a man could be measured by how much 'leisure' his wife had. The more prosperous a man became the more likely it was that his wife would not undertake waged labour. However, these women did of course work in maintaining households, supervising staff and establishing and maintaining social contacts for the family. Women also continued child-related work.

The result of industrialization was an increased separation of home and work spheres; the development of dependence on wages for survival; a gradual entrapment of women within the home area; and a gradual devaluing of women's traditional skills. The latter were taken over by the development of, for example, the textile industry. As capitalism developed, the labour of production and reproduction were split.

Men's role under capitalism is to labour and produce within an hierarchical ordering of worth, to develop surplus goods to be sold for capital. Those who own the means of production (state or individual or corporation) own the products, therefore surplus, therefore capital. Woman's role is to reproduce the next generation of workers, but also to service men to continue within the workforce. Her role is the servicing of capitalism, through her responsibility for home/family maintenance, domestic chores and child socialization. This is all unwaged labour and serves both men and capital.

Women became the dependents of men in a new economic way and 'this dependency had become embedded in the workings of the larger economy'.[20] This financial dependence also led to the push for a 'family wage' for *men* (and only men), and women's need to marry well.

At the end of the nineteenth century, with increased technology and industrialization, men's wages increased. Some positions also began to open up under pressure from women. The 'family consumer economy' developed 'as households specialized in reproduction and consumption'.[21] (The role of women as consumers will be discussed in the section on houseworking.)

But the allocation of domestic work to women and its concomitant devaluing by men did not come about solely because of this increased separation of home and workplace. Patriarchal ideology in the family led men to ensure that workforce changes would empower them further within the family, relative to women. A gradual diminishment of the extended family to the nuclear family took place. With it came the middle-class ideology of privacy within the family and the growing isolation of women from other women kin. These changes were also essential for the development of capitalism, which relies on high consumption by small units. (For example, *each* family must have a lawn mower even though it is used once every three weeks, and could easily and more profitably for the individual family—but not the manufacturers—be shared between three or four families.)

Hartmann points out that often discussions of 'the family' as a social entity have assumed a unity of interests within it, presenting it as an active agent but neglecting conflicts of interests or differences within it. But the family is a 'locus of struggle'. The same processes (industrialization, for example) which established households as partners with, but in opposition to, the state, also augmented the power of men within households, increasing tensions within them.[22] The institution of the family still defined the work roles of women and men. 'Values, behaviour and strategies shaped under one mode of production continued to influence behaviour as the economy changed. The older practices were adapted to the new circumstances'.[23] The family was deprived of its productive function as an economic unit, but women were particularly affected as the movement of men into waged labour gave them greater power over women's unwaged labour. Family ideology continued to influence the work activities of its members and that ideology was male-dominated and controlled. In pre-industrial society male ownership of women's labour was well established. Middleton comments that:

integrated and independent these household economies may have been; but to suggest that they were founded on a marital *partnership* is to invest language with an uncommon degree of flexibility. Authority, in fact, was strictly patriarchal. Essential powers of decision-taking, especially in regard to the disposition of resources, resided with the household-head.[24]

Women's labour is appropriated within the marriage contract by men. Christine Delphy calls this the domestic mode of production. It includes the consumption and circulation of goods, but, in comparison to the exploitation of workers under capitalism, those who are 'exploited by the domestic mode of production are not *paid* but rather *maintained*. So consumption is *not* separate from production and the unequal sharing of goods is not mediated by money'.[25]

Barrett argues that the 'family', which Rowbotham says is 'maintained at the expense of women',[26] is not only the means of maintaining capitalism, but also a 'product of historical struggle between men and women'.[27] The patriarchal ideology which confines women to the family and justifies the sexual division of labour is necessary to capitalism. Thus, as Hartmann points out, the *material* base of patriarchy is men's control of women's labour both within and outside the household.[28] Male-controlled trade unions, for example, have denied skilled jobs to women, and the denial of abortion and child care reinforces women's financial dependence on men and therefore on their waged labour.

In their analysis of work Game and Pringle illustrate different aspects of the relationship between gender, the labour process and technological change in the current context. They show that masculinity and femininity are socially constructed *in relation* to each other and so also are women's and men's jobs. They state that 'gender is fundamental to the way work is organized; and work is central in the social construction of gender'.

Gender is not just about *difference* but about *power*, 'the domination of men and the subordination of women'.[29] Thus men's work must be experienced as empowering, and acceptance or rejection of changes within the workforce depends on whether or not they will diminish the male worker's sense of prestige and power. To this end, working-class men also 'have agency in preserving male power'. Sexuality is also part of the male domination of women and expresses itself in the workplace through, for example, men's use of sexual harassment (though also in more subtle and integral ways).[30]

Patriarchy, which gives some men power over others, but all men power over women, is integral to explanations of change, or the lack of it, in the workplace. As Game and Pringle write:

Men at the bottom of the male hierarchy still fight to retain this power. Those at the top often assert it even when it seems to be in contradiction to their economic interests. Just as they will often forego surplus value if necessary to increase their control over the labour process, they will forego it also to maintain the sexual division of labour. It is in this sense that we can say that the inner logic of capital is patriarchal. Capitalist rationality is based on male dominance.[31]

They discuss three forms of control over the sex-segregated labour process. The first is direct and personal, relying on the power of the individual, for example, the entrepreneur. It is characterized by 'family' symbolism, so that the controlling male represents the 'father' of the business. Women's position is seen as synonymous with their place within the family. The second is technical control, where masculinity is designed into the machine and the technology actually paces the work. Here men use the big machines and women the small. The third is bureaucratic control, where male dominance is part of the social organization through rules, procedures, job descriptions and evaluations. This control *seems* fairer to women in its supposed objectivity. But, in fact, the impersonal indifference of bureaucracy, established with male as the norm, allows rationalization of the position of men. For example, it ignores the specific problems faced by women workers.

Thus, Game and Pringle argue that the sexual division of labour cannot be understood in purely economic terms. The sexual and the symbolic are crucial in the interaction, and labour issues must also be fought at that level. Barrett also makes this point: that the oppression of the workplace occurs within capital, within materialism, but also within ideology.[32]

The role of the state is also an important component. Hartmann suggests that the state developed by breaking the power of kin groups and promoting households with male heads. The power of men was thus enhanced and related to the contruction of household units.[33] McIntosh also argues that the state oppresses women through its support of a family household dependent on a male wage and female domestic servicing. 'This household system is in turn related to capitalist production in that it serves (though inadequately) for the reproduction of the working class and for the maintenance of women as a reserve army of labour, low-paid when they are in jobs and often unemployed'.[34] Women's oppression through work thus originates in an interaction of class, family and state, but patriarchy is at the core of it, because ultimately it is men who benefit from its maintenance.

The patterns of women's work are also controlled by men. Tilly and Scott point out that 'the interplay between a society's productive and reproductive systems within the household influences the *supply* of women available for work. The characteristics of the economy and its mode of production, scale of organization, and technology influence the *demand* for women as workers'.[35] But in 'demand' situations a primary constraint operates—the universal segregation of jobs by sex.

Occupational segregation by sex

The current statistics on women in the waged workforce show patterns which are similar across the United Kingdom, Australia and the USA. Women have increased their presence in the acknowledged paid workforce. Representation of women in the labour force has been increasing at a more rapid rate than male participation.[36] This is particularly so for married women, whose participation rate more than quadrupled between 1947 and 1966 and has climbed further since.[37]

Australian figures published in April 1984 show that the participation rate (that is, proportion of the female labour force wanting work or working) for women over fifteen years of age was 47.3 per cent. Employed women made up 38.9 per cent of the work force. Women were 29.6 per cent of the full-time employed and 78.9 per cent of part-time employees.

The participation rate for married women with dependent children (as of July 1982) was 45.7 per cent; for those without dependent children, 38.5 per cent. Thus, married women *with* dependent children were slightly more likely than those without to be in the labour force. Women were more likely to work part-time if their children were dependent. Women who were heads of households with dependent children were less likely to work at all, particularly full-time. 'For both women in married couple families and female heads of other families, it appears that dependent children were not a deterrent to women's participation in the labour market but that they affected the number of hours women worked'.[38]

These figures did not yield Aboriginal women's employment statistics. But migrant women (those born outside Australia) made up 25.4 per cent of the total female labour force.[39] In the United States a greater percentage of black than white women work,[40] although the gap is closing. A primary reason for racial differences in participation rates is the greater number of women of colour who are single heads of households and in need of the money.[41]

Information on the class affiliation of women working is difficult to glean, but some information on women in relation to husband's income is obtainable. Blau has indicated that wives are more likely to work if the husband's income is in the middle range rather than high or low, and Stroker points out that this increases the gap between the poor and middle-income families. However a paper by O'Loughlin and Cass on Australian data indicated that wives with husbands in middle-income or very high income brackets had lower participation rates. The differences seem not to be striking however. O'Loughlin and Cass state that 'It is clear that the wives of men earning below average weekly earnings have the highest labour force participation rate (67.4 per cent), followed by the wives of men earning about average weekly earnings (62.4 per cent), while the wives of men earning above average weekly earnings have the lowest participation rate (58.6 per cent)'.[42] Wives with husbands on higher incomes are likely to seek part-time work. So women are more likely to be employed and employed full-time if husbands are low income earners and the correlation is strongest when there is a dependent child in the family.

An outstanding characteristic of women's workforce participation is the occupational segregation of jobs. Women's traditional household work is extended into the market place. So there is a dominance of women within service areas of work, often called 'pink collar' jobs (as opposed to the blue/white collar work of men). Again figures are similar for North America, England and Australia,[43] though a comparative study by the OECD revealed that in 1977 Australia had the highest degree of occupational segregation of the sexes.[44] The majority (59.5 per cent) of working women are in service, sales and clerical positions. These are female-dominated jobs, for example 73.8 per cent of the employees in clerical work are women. However the Women's Bureau points out that 41 per cent of these women worked part-time as opposed to 12 per cent of men. Of the 19.1 per cent of women in professional and technical occupations, 37.8 per cent were teachers and 30 per cent were in nursing, again traditional female occupations. Only 2.5 per cent of women worked in administrative, executive and managerial positions, compared with 9.3 per cent of men.[45] These figures support Berch's division of segregation into *vertical* segregation— *within* jobs with men at the top, and *horizontal* segregation—the disproportionate representation of women in certain jobs and an absence in others.[46]

Sex segregation of jobs is universal; however, differences in the particular jobs assigned to men and women can occur in different cultures. For example, medicine is an acceptable job for women in the Soviet Union, so women predominate; but they are mainly general practitioners who are paid less than bus drivers. Surgeons, paid very well, are predominantly male.[47] This division is socially structured for a particular purpose. Greater importance and value is accorded to the jobs which men perform.[48]

Because men's work is structured to empower them in relation to women, changes in the sex labelling of jobs is resisted as it affects this relationship, unless the status of the job itself is falling.

For example, clerical work is seen now as women's domain, yet when businesses were small in the nineteenth century clerical work was semi-managerial and mostly done by men.[49] 'Skill', in the concept of 'skilled work' is a gender-biased term. Skilled work is men's work, and women's skills are not recognized. This was an important distinction once used by the trade unions to maintain male domains and therefore higher male wages. But what about the capacities needed in women's jobs, for example coping with monotony, boredom and the pressure of factory work? Service jobs do require skills; the traditional female skills of interpersonal interaction.

The structure of the workplace reinforces segregation, based as it is on the assumption that workers will be men. It assumes, for example, back-up domestic support systems (a wife), and neglects the need for child care, leave for illness of a child, or fair payment and terms of employment for part-time work. Traditional attitudes to work, based on a false assumption that women and children have a willing and able breadwinner within the household, support the male-oriented structures of the workplace. For example, if we look at a career position for a woman rather than a job, women are greatly disadvantaged by the assumption that the child-bearer is solely responsible for that child and domestic duties. Most professions require flexibility to attend meetings after work, to take extra work home, or to travel to conferences. This path to success is only possible with family support, few domestic chores, and extra time, all of which are generally unavailable to married women with children yet are available to married men through the unpaid domestic labour of their wives.

Game and Pringle argue that the sexual division of labour is maintained through an acceptance and constant reproduction of gender difference within the workplace. Men use the sex-labelling of jobs to reinforce the power relationship between the sexes. A *distinction* must be maintained, even if the jobs so labelled change. One example is the banking industry, in which teller positions originally 'belonged' to men. In the 1950s women were allowed into these positions at a time when mechanization took place. But management positions were reserved for male recruitment.

They hoped that women would do the menial jobs and then leave after five years, thus reducing the pressure for promotion. In this way careers were saved for men although the bulk of bank work was being performed by women. A middle-class masculinity was being preserved through a sexual division of labour which disguised fundamental changes in the organisation of bank work.[50]

The increase in the number of women working part-time in the labour force skews women's apparent improvement in terms of their workforce participation. The growth in part-time jobs accounted for four-fifths of the rise in women's employment between 1974 and 1982. Most married women in part-time jobs do not want to work more hours. Among unemployed women, 82 per cent of married and 52 per cent of other women want part-time work. Eccles points out that this 'preference' must be interpreted 'in the context of the severe shortage of child care and the present social structure under which women take the main responsibility for unpaid work in the home'.[51]

Part-time working conditions are poorer than those in full-time work, with fewer non-wage benefits and little chance of training or advancement. Employers have fewer obligations to part-time employees, who are also easier to sack. Women's willingness to work part-time allows avoidance of a real restructuring of the workplace itself to suit the needs of full-time women workers. Women are maintained as a floating labour force which can be increased or decreased when 'appropriate' in the male-run economy.

Women's workforce participation is manipulated in other ways too. The experience of women during World War II illustrates some of these. The first is the total exclusion of women from some areas of work by legislation, for example, within the armed services or heavy industry, based on a clear distinction of 'appropriate' work for the sexes. Women were used in 'men's' jobs during the war, while men were assured that their traditional employment opportunities would not be threatened. Prime Minister John Curtin announced in 1942 that 'all women employed under the conditions approved shall be employed only for the duration of the war and shall be replaced by men as they become available'.[52]

Women's low pay rates were (and still are) a disincentive to entering the workforce. Before the war, women's basic rate of pay in Australia was set at 43 per cent of the male rate, so women needed to be very keen or very needy to look for work. With a lack of union support, conditions were also appalling. Curlewis documents the depressing and debilitating conditions faced by women conscripted into industry. Rates of absenteeism were high, often due to family responsibilities but often due to ill health.[53]

The availability of work for women is also manipulated by making available or withdrawing services to assist them in their domestic responsibilities, such as child care. War workers, for example, were provided with creche and child-care facilities, as well as receiving a good deal of film propaganda encouraging women to use them because women's work was essential to the war effort.

After the war, ideological manipulation ensured that a different message was conveyed to women. The 1983 film *For Love Or Money* showed the constant propaganda admonishing women to return to the home to care for the traumatized men and children of the war years. Their new job was to create a haven of peace. As Beaton writes: 'the whole burden of domestic and psychological rehabilitation was thrown on to women'.[54] Moreover, the proximity of the Japanese had impressed upon Australians their position as an isolated white outpost within Asia and women were encouraged to populate the country.

Many women did return to the home as 'requested'. Many were just sacked. But statistics indicate that there was not the mass exodus of women from the workforce which we have been encouraged to believe. The pressure did work on some women, but the majority stayed within the workforce. Beaton points out that the misleading picture of a mass exodus is dangerous in that it has 'helped to re-establish the invisibility of the working woman'.[55] It can also be used now to give women a self-image of the easily manipulated docile worker; a tradition of women working when needed and giving up when the men decree it. If this self-image is accepted as 'the way women are' or should be, it encourages further manipulation of women out of the workforce in times of economic stress. It increases the psychological pressure placed on women through guilt if they work in times of recession.

Ideological pressure on women continued throughout the 1950s in a new guise. Child-care 'experts' began to create literature on 'maternal deprivation'. Workforce mothers came under constant pressure to justify their choice to seek paid work. John Bowlby asserted that maternal absence, even for short periods, could have devastating effects on a child. These studies were based primarily on institutionalized children and showed in reality the impact of institutionalization, not of a mother's absence. But the image of 'latch-key kids' becoming vandals and unbalanced adults put pressure on women to stay at home. The message was, if you leave your children to work, 'do not expect them to thrive'.[56]

Research on working mothers indicates that the most satisfactory mothering comes from women who are satisfied and content, whether it be at home or at work, and that children of paid working women have the advantage of learning an independence early in life which prepares them for self-sufficiency in adulthood.[57] Only recently has the problem of the constantly absent father been a focus in discussions of child-rearing.[58]

Women are also a major part of the subterranean economy, which consists of those people working 'illegally', that is not paying tax. Mostly this involves traditional women's services like cleaning houses and baby-sitting, or working in family businesses (usually unpaid). Berch notes that according to 1980 figures, 'the total value of economic activity in the US subterranean economy was around $7000 billion per year, or about 27 per cent of GNP'.[59] This labour is largely unrecognized.

But the most exploited women workers are the pieceworkers, women who take in work at home for factories of various types. Juliet Mitchell has commented on the appalling situation in England in the 1960s and the 1970s but the current situation in Australia is no better.[60] Simons estimates that in 1983 there was a hidden workforce of pieceworkers of 30 000 people, mostly migrant women. They are cheap labour for clothing and textile factories, where the employer pays no overheads and can set the pay and the limits which piece-workers have to reach. One, for example, made women's briefs at the rate of 120 pairs an hour, and was paid 20¢ per dozen. Each pair was sold in supermarkets for $2.99. The Centre for Urban Research and Action notes one Turkish woman who made evening dresses for $2.70 each in her garage and they were then sold for $70 in the shops.

Pieceworkers pay for their own electricity, machines and maintenance. Some are members of the Clothing and Allied Trades Union, though only 1 in 150 are. They are isolated and exploited workers and threaten unionized factory work. Yet the unions face a quandary over them: if they take action on behalf of pieceworkers, the women may just lose the work. Piecework takes advantage of those women who are poor and need the money, but who are unable to leave their homes for work because they have dependent children.[61]

Some women, of course, cannot get paid work. Women already have a higher rate of unemployment and figures indicate that the gap between women and men is widening.[62] In Australia in October 1986, of the people not in the workforce who wanted a job 77.4 per cent were women and 22.6 per cent were men. Women also predominate among discouraged job-seekers. In October 1986 a group of 237 200 Australian women not in the labour force gave family considerations as their reason for not actively seeking work; 29.3 per cent of these cited lack of suitable child-care; and 49.0 per cent cited the desire to look after their own children.[63]

Wage differentials

In North America, England and Australia, women still earn less than men. The term 'equal pay' continues to be misleading, as the average weekly earnings of women in Australia still represent only 66 per cent of male earnings.[64] In Australia this difference between male and female wages has been legislated into the system because 'the practice of wage discrimination against women workers was institutionalised at the turn of the century'.[65]

This occurred with the 1907 Harvester judgement delivered by Justice Higgins of the Federal Arbitration Court, when the concept of a 'living wage' was crystallized and remained within wage-setting debates and procedures for the next sixty years. The 'family' or 'living' wage was based on the assumption that family financial responsibilities would be borne by men. Women, it was assumed, would or should have a man to support them in return for domestic services.

The basic rate of wages in Australia was strongly influenced by these assumptions. In 1912, Higgins fixed a single-unit female wage at 60 per cent of the male 'family-unit' wage in the Female Basic Wage Case.[66] In the clothing trades in 1919 the female basic rate was set at 54 per cent of the male rate. The Basic Wage Inquiry of 1949–50 determined the rate at 75 per cent.[67]

The battle for equal pay for women in Australia has been constant and long, with at times a devastating effect on women. Women's position in the Depression for example, was severely affected by the assumption that there was a male breadwinner for every woman. Many women were denied sustenance funds, and many single women became destitute. In 1932, W. Kent Hughes was made Minister for Sustenance in Victoria and declared that 'as long as domestic work was available—at *any* wage, under *any* conditions, *anywhere* in Victoria—unemployed women would be denied public assistance'.[68] The government worked hard to cut women from the relief rolls, moving the ratio of men to women seeking relief from 19:1 in 1932 to 140:1 in 1935.

Marchant has pointed out that there was 'a systematic attempt to certify women's inferior position through wage fixation'.[69] This penalized any woman who was single or who had dependants, and advantaged single men who still earned a family wage.

Various explanations for the institutionalization of inequality with respect to wages have been proffered. The 'human capital' or supply explanation contends that women earn less because they are inferior to men, as indicated by level of education, experience, work continuity and so on; that is, they earn less because they are worth less. A second explanation holds that employer discrimination operates: employers refuse to hire women in the higher-paying jobs, for example in managerial positions. The third 'crowding' explanation holds that because employers discriminate in the higher wage brackets, women are crowded into the less desirable and less well paid jobs.[70]

These explanations contribute to the picture. Women are less well educated and work less continuously within the workforce in Australia because of expectations about their social role in the domestic sphere. But where education is equal, women are still paid less. Chapman writes that 'disaggregated empirical investigations reveal that apparently identically qualified women and men receive different remuneration, even within narrowly defined job categories'.[71] Criticisms of women's work performance based on continuity of service or lack of personal investment in a job are grossly misleading when the male-centred structure of these criteria are analysed. Child-related absenteeism, or potential absenteeism, on the part of women becomes a ruse for refusing them jobs with increased responsibilities.

Pay inequality also operates within particular organizations where women and men have equivalent training and experience. Analysis of the causes of these inequalities led Blau to conclude that discrimination accounts for a significant share of the pay differential. After all factors which might tend to cause productivity differences between women and men are controlled, 'the proportion of the sex differential attributed to pure discrimination has been estimated at between 29 and 43 per cent of male earnings'.[72] It is maintained because both men and capital gain from the situation and it is continued through the setting of criteria for advancement which are male-centred.

Analysing the debates on the origins of the sexual devision of labour, Curthoys lists a number of variables. As indicated above, these include the fact that economic support comes through wages but only part of the population directly earns them. Private caring, including child care, remains within the family, where those dependent on the wage-earner do the tasks unpaid. Women primarily do this work. Women's participation rates in paid labour are lower than men's and they are paid less. Jobs are segregated by sex and paid accordingly. Once a job is sex-labelled, women and men are differentially trained.[73] And so the system is perpetuated, to the benefit of man.

Why women seek waged labour

Pay inequality continues to operate; jobs are difficult to find and will locate women in the less desirable areas of the workforce; and women have the 'excuse' of domestic ideology for staying at home. So why do women seek paid work?

If the question were asked of men it would seem ridiculous and the answer would be 'to earn a living of course'. Well, as Berch says, 'like everyone else, women work to eat'.[74] Two-thirds of women workers in North America are single, divorced, widowed or separated, or have husbands who earn low incomes. Married women who work full-time are contributing on the average 40 per cent of the family income.[75]

In Australia since 1966 there has been a doubling of the proportion of one-parent families headed by a woman. O'Loughlin and Cass indicated in 1984 that 84 per cent of these families are dependent on social security and a third were living in poverty.[76]

The situation is similar in the US and England and is creating the feminization of poverty, or more properly, the pauperization of women. Many of these women are unable to work because no facilities are available for the care of their children, workplaces being based on the needs of men with the assumption of female support. Trends indicate also that fewer women are now marrying. So in essence, many women work because they need to support themselves and (often) dependants. Others work because the survival of the family unit depends on it. For example in 1971 in the US 14 per cent of all two-parent families would have been classified as poor if the wife had not worked.[77]

Women may also work for economic reasons not related to survival. Family aspirations may require added income to boost the standard of living. This may be measured by the acquisition of a car, colour television, added education for the children. It may also include so-called 'luxury' goods, which the society impresses upon people as *socially* (as opposed to *physically*) necessary. Social expectations play a role in what is seen as 'necessary' within the family unit. Leghorn and Parker comment that it is 'overwhelmingly women's work which makes the poor and working-class families able to meet the culturally defined minimum standards'.[78]

Women also want waged labour because work that is paid for is valued and recognized by society. It satisfies a search for meaning and for a position in the world. A job 'fixes one's place in economics, politics and society, while influencing friendships, family relations and even types of recreation'.[79] It brings order into life and structures the day. Though it may isolate a person from home relationships for working hours it can increase the value of the time spent within those relationships. It contributes to a person's sense of identity, self-esteem and status.

A job brings social contacts with it. It gives a woman a social role apart from that of wife and mother. She is, in addition, a 'worker' with working power and a set of people who relate to her differently from family members. In a study of unemployed women, the New South Wales Council for Social Services found that even when women worked because they needed the money, their satisfactions through work were not limited to financial gains. Like men, paid work gave them occupation, a 'purpose and identity through which they define themselves'.[80] Most of the women interviewed had been working in relatively unskilled jobs, but preferred them to staying at home. Because labour under capitalism has been seen as alienating it is often assumed to be individually alienating. Yet the evidence of this and other studies indicates that most paid work carries with it certain intrinsic satisfactions over and above the money involved.

One motivation for working is to obtain an equitable distribution of income within the marriage. Although wives often manage household finances, *control* of them is much more often in the hands of husbands. Here 'management' is similar to that in firms, where it means the carrying out of decisions made elsewhere. Control refers to overall power to decide how money is to be spent. The controller naturally determines who will benefit most from family income. In one Australian study, distribution was in some cases so inequitable that some wives had a lower standard of living than their husbands. Women without their own earnings were the most vulnerable because of their dependence on the husband's income and the need to *ask* him for extra money when necessary.[81]

Women want paid work in order to exercise some control and power within their own lives. Earning power does give a woman increased rights in society and within marriage. A wife's work and educational experiences influence her marital power. Gillespie found that 'regardless of class, women working outside the home have more power than those who do not. And the longer they have been in the labour force, the greater their power. They have the right to be active in decisions made about moving the family for a job change and in the spending of family income'.[82] Many women only recognize this when they stop earning after the birth of their first child.

In summary, women work for the social contact it brings, for benefits in self-identity and esteem, because of financial need, and from a desire to retain some control in decisions made affecting their lives.

Double lives

The emotional tension of balancing family and work roles often brings women to the point of exhaustion. The fatigue experienced by working wives and mothers, coupled with guilt and conflict, can result in ill health. Even in 1944 Doris Beeby, organizer of the Sheet Metal Workers Union in New South Wales, noted that a number of women members had broken down under the strain of factory work plus domestic responsibilities.[83] Men remain blind to this dual workload as long as the family functions well.

In her detailed analysis of studies on housework labour in North America in the 1970s, Hartmann concludes that husbands of waged workers do not do more housework than husbands of non-waged workers. Employed wives spent more time on housework on their days off; while husbands use their time for leisure. Working wives spend less time on house duties than do non-waged wives, but their total working week is longer. When young children are in the family, studies indicate that the wife's work week expands to meet the needs of the family while the husband's does not.[84]

Technology has not reduced the amount of time spent on household duties. Machines may have reduced the amount of effort for each chore and permitted them to be done at less specific times (for example, tumble driers enable women to fit in washing after work). But as standards of cleanliness and nutrition have risen, so too has the time spent in attaining them. Women are responsible for the servicing, cleaning and repairing of gadgets. Men increasingly relinquish their traditional work. Lighter lawnmowers mean that 'liberated' women mow lawns. They do all manner of home repairs. Game and Pringle write: 'Earlier in the century, gas and electricity relieved *him* of the burden of chopping wood. When *she* is given a rubbish disposal unit and a dishwasher, he relinquishes any responsibility for washing up or putting out the rubbish while she gains the extra servicing work'.[85]

Harper and Richards found in Australia in 1979 that although husbands tend to help more with child care they have not increased their household chore workload. They also found a variation across class with respect to the willingness of husbands to help with household chores. Husbands of professional workers tend to do more housework than those of non-professional workers.[86] Surveying the evidence from other countries, Bryson concludes that in 'most cases women still carry a heavier total burden of work (or overload) in terms of daily and weekly hours than their husbands'.

Bryson summarizes the findings on Australian marriages and draws three conclusions. First, the distribution of domestic tasks has changed little over the last thirty to forty years. Although men have increased their participation in household tasks, women still give a significantly greater contribution. Second, some fathers are taking a greater role in parenting than before. And third, 'women with higher levels of education are more successful at extracting contributions from their husbands than are those with lower levels of education'.[87]

Some writers express uneasiness about the increased power of men within the domestic sphere. Game and Pringle point out that women may be reluctant to give up the responsibility of household management because: 'not only is it the site of their oppression but also a space defined as theirs to exercise some control over'.[88] Bryson comments that while women may have dreary chores reduced, they may also lose organizational control and 'indeed become more subject to control by their spouse'. She questions whether increased home labour and fathering in particular leads to gender equality, writing that it may result in 'a net loss of power and control' for women.[89] A sense of identity needs to be located in some centre of personal power, and for many women this is home-based. The question is, can the initiation of men into equal sharing of 'chores' relieve women's burden without making them less powerful?

Women also experience stress because of the clash of values which seem to be required within home and workplace. This is particularly true of professional career women. Bernard has discussed the ethos of women's worlds as being one of love and duty, while the ethos of the male world is one of self-interest. The public sphere strongly represents male values. To be successful women are required to use the guiles of stereotyped femininity or overtly challenge the male sense of power. Trained to be compliant and agreeable, women often find it difficult to enter the competitive world of the economy as it is now structured. How to succeed within this world and keep the positive values of caring and empathy is problematic. Committed to group, rather than individual goals, women often try to change the processes and operation of power and to create democratic decision-making.[90]

Houseworkers

Women may be 'housewives'[91] or houseworkers for a number of reasons: they cannot get paid labour; they can afford to work at home financially and prefer it; they want the opportunity to spend more time with their children when they are younger; or they are so exhausted by the dual workload that they prefer to be doing only one job!

Housework is a multi-faceted job, involving service tasks in a variety of areas, and management skills. It includes the servicing of children and men through feeding, clothing, cleaning and so on, as well as the organization of the social arrangements of the family and its connections, for example with friends and relatives. The features of the role (e.g. that it is supposed to be constant servicing), as well as the social invisibility of the work, conspire to conceal the amount of time it takes. The work is private, isolated, self-defined: men do not see it being done because the routine tasks result in 'the creation of a normal environment for him'.[92] The private household is meant to 'serve as a counter-balance to alienated work in the public sphere'.[93]

The houseworker meets the *consumption* needs of the family unit, both in the purchasing role and as the 'facilitator' of the consumption of others (e.g. as hostess). Housework no longer demands as much time in the production of goods for the family, for example soap, candles, bread and so on, so the role has shifted to one primarily of *consumption*, which includes not just the 'using up' of commodities, but their acquisition and transformation. Purchasing is now a major job of the houseworker. Her value is reflected in her success as a shopper. This involves research, comparison shopping, and often bulk buying, which is very time-consuming. Because advertising is so closely related to packaged sexuality, the process also becomes 'substantially about the purchase of love and approval and the construction of an appropriate self-image'.[94] To be a 'good wife' is now to be a 'good consumer'.

The tasks of emotional labour become conflated with the tasks of house labour, imbuing servicing work with an added dimension. Doing the washing thus becomes a 'labour of love' by women for the family, instead of 'work'. Home-making is tied to family loyalty, which means it is seen as the things women do for those they love. This convenient ideology relabels work and means that to complain of the *work* is to complain of the family, husband and children.

Domestic chores are seen as non-productive, so in the male definition of work as 'productive' women's labour is made invisible. The chores themselves are dull, repetitive, mundane, and involve cleaning and servicing for others. This kind of work, when it is unpaid and unacknowledged, is 'directly opposed to the possibility of human self-actualisation'.[95] Oakley points out that the responsibilities of the job *are* motivating, but the reality of the work is not, and motivation in a job is a primary source of fulfilment. The job is not one from which the worker can gain any sense of herself or any information about herself as a person with meaningful work to increase self-esteem.

There have been attempts to estimate the economic value of women's work in the home. These have been motivated by women themselves claiming acknowledgement and payment for home labour and by insurance companies encouraging men to insure their wives. There are a number of methods by which home work can be valued.

The first is the 'market cost approach'. This tries to estimate the services as they would be paid for in the market place, for example laundry, chauffeur, cleaner, administrator and cook are priced. The 'replacement cost approach' is a method which estimates how much it would cost the family to replace the wife or homemaker. This method provides a better estimate of the value of the wife to the family. But a third approach places the woman at the centre. The focus is on what it costs the woman in earnings foregone to stay at home: the 'opportunity cost' method.[96]

Estimates are difficult to arrive at because of the difficulty of measuring the tasks done. These are often carried out simultaneously, for example planning the shopping list and evening meal while driving the children to school. But American insurance companies are now paying up to $15000 a year, depending on the household.[97] Leghorn and Parker cite figures of US$257.53 and US$700.00 a week in value. In North America if women were paid for their home labour based on a 1970 survey, it would be 'more than half the Gross National Product, and five to six times the military budget of the mid-70s'.[98] But once these estimates started appearing, men again claimed that it would be 'ridiculous' to consider paying for women's 'natural work' done out of love. And, after all, with all the time-saving devices available, women really had leisure time, not work.

The push for payment for housework grew in the 1970s, particularly among conservative women's groups. Various methods were suggested, including the joint payment of a husband's pay to the couple, or subsidizing it by government payment. These have been rejected by capital, because it relies on this unpaid labour, and by men, who benefit from it and from not seeing it.

Oakley, however, has argued that if it were paid, the woman-equals-housewife equation would become institutionalized. The woman would not have the benefits of a *group* of workers and would still be isolated and with a negative self-definition. The wage would be like a grant. It would create a new form of oppression. And certainly if housewives were paid, more women could be coerced *out* of the paid labour force on the grounds that they could be paid for domestic work. Oakley comments that: 'its character would be that of a benevolent handout, rather than a payment earned of a *right* for the work done'.[99]

The 'wages' of full-time housework have a negative effect on a woman's self-identity. Feelings of low self-esteem and resentment of financial dependence often occur, as well as a sense of loss of control and meaning due to the repetitive nature of the tasks which are done, only to be redone. In 1956 Myrdal and Klein listed clearly the problems of the houseworker, and these remain unchanged. They are: houseworkers have the status of a minor through financial dependence; social isolation is prevalent; they lack mental stimulation (which children do not provide); they are victims of the ideology of privacy; and they lack the satisfactions of 'team work'. In addition, marriages are often strained because the husband may be the sole source of a woman's intellectual and spiritual satisfaction, as he forms her basic and sometimes only link with the world at large. Husband and wife 'live in two different worlds and the area in which the two spheres overlap has been emotionally overburdened by the social isolation of the housewife'.[100]

Rowbotham has pointed out that when a woman sinks her 'identity into someone else you suddenly get a terrifying feeling that you are no longer there'[101] Together with an unstructured day, which many people find difficult to cope with, these problems can lead to depression and a variety of effects usually labelled 'nervous disorders'. Alcohol and drug abuse, aided by the over-prescription of anti-depressants by an unsympathetic medical profession, are on the increase. In general, houseworkers have been found to be poorer in mental and physical health than other women.[102]

To cope with these problems, women have devised methods of resistance within the frameworks of their lives. Strategies include 'switching off, the half-there swimming feeling, the barriers around yourself, and then there is illness'. To feel indispensable is important to all individuals, and women have attempted to value housework and find happiness within it. Like all subordinate groups, Rowbotham maintains, women have developed an 'alternative myth of self-respect'.[103]

Housework thus devalues a woman's self, draining her identity as a person of worth. Leghorn and Parker comment that ultimately 'women's burden comes not from their lack of access to man's world, but to the separation and division of power, benefits and responsibility between men and women, which is rooted in women's role and work in the home'.[104]

Further issues

The issues raised here in relation to both women's self-identity and work, and their oppression as a social group, are general points. The particular ways in which women are constrained through the construction of the 'work world' is being continually elaborated in detail. This exploration includes the individual stories of women as they experience paid labour in a variety of industries, theoretical analyses on the construction of work, and studies of specific industries and how they function in women's oppression. A number of important issues are arising from this work, among them sexual harassment, technology and its impact, unionism, and child-care issues.

There is not space here to deal with these issues in detail. Briefly, the work on sexual harassment indicates that the gender construction of the workplace reinforces a dominant role for men, while continuing to make women the 'other', intruders within the public domain, accepted on sufferance. The sexual oppression of women is continuous throughout private and public spheres. From the sexual abuse of women through advertising to the assumption by male workers that they can touch women workers and enter their personal space uninvited, the basic partriarchal ideology makes women objects to be handled, used and disposed of at the whim of men. The difficulty incurred in 'proving' either a case of discrimination or one of sexual harassment is that there is no dividing line between 'normal' heterosexual behaviour and sexual harassment or abuse. The fact that men also have within the workforce added authority and power through status and occupational segregation makes harassment easier.

Studies of technology currently revolve around health hazards and replacement hazards. The first has become particularly obvious in the areas of clerical and factory work, where repetition strain injuries are becoming frighteningly more common. These injuries of the muscles and tendons, including tenosynovitis, carpal tunnel syndrome, tendonitis and neck and shoulder pain, are caused primarily by rapid repetitive movements demanded in various occupations, particularly the occupations dominated by women workers.[105] In 1981–82 in New South Wales, for example, in the category synovitis, bursitis and tenosynovitis, 295 new clerical and professional compensation cases were reported, and 273 of them were women.[106] There is also concern over the use of visual display units (VDUs) which include word processors, frequently used now by primarily female secretarial workers.

In an excellent overview of technology, work and women's health, Thompson and Neary of the Women's Bureau, Federal Department of Employment and Industrial Relations, outlined ways of categorizing

the occupational health problems of working women. These are: problems associated with both men and women in similar jobs; those related to 'men's' jobs but into which women are moving; those related to 'women's work' jobs; and those which affect fertility and reproductive functioning, where both sexes may be affected but the foetuses of pregnant women are particularly at risk.[107]
Petchetsky has dealt with the politics involved in making decisions about reproductive hazards. She cites a case where women chemical workers at American Cyanamid's Willow Island Plant were given the 'choice' of giving up relatively well paid 'non-traditional' production jobs or being sterilized. Some chose sterilization. Unfortunately, the company folded soon after, leaving those women sterile and unemployed.[108] Irving cites a case in Canada in a General Motors battery plant, where only sterilized or non-fertile women were permitted to work.[109] These are only two of many examples. These decisions are made regardless of the fact that men's reproductive functioning would also be affected by the chemicals or radiation. By eliminating women workers or forcing sterilization, management controls women and can exclude them, while failing to protect male employees.
Petchetsky points out that 'protective' legislation aims to exclude fertile women from working in specified areas, which can be defined and redefined according to the legislators. There are three principal arguments against this exclusionary 'protective policy'. First, many traditional female jobs, for example hospital work, have reproductive risks involved, yet they do not exclude women workers; the 'protections' being offered to women are selective and arbitrary. Second, the treatment of women as 'potential mothers' ignores that 'motherhood is a social relation, not a biological status'; men are not excluded either, regardless of the reproductive dangers to them as 'potential fathers'. Third, 'protective' legislation should be just that: it should be protective for all workers and none should be exposed to such hazards.[110] Essentially then, protective legislation has been used to define women in terms of biological roles and to exclude them from certain jobs. At the same time it ignores the hazards in many other jobs which women do.
There is a range of work-related dangers. These include physical danger (for bank tellers, for example), sexual harassment or physical threat; and possible health problems, for example librarians and archivists are exposed to dust and spores conducive to bronchial diseases and allergies.[111]
Thompson and Neary claim that the reasons why women's health issues are ignored are rooted in occupational segregation. They write that 'the continued practice of piecework and bonus systems, the fatigue caused by the introduction of electronic keyboard equipment and visual display units, the dual role of worker and mother and discriminatory protective legislation have not assisted in the development of a healthier working environment for women'.[112] The wide range of hazards for women is rarely clarified.

Another technological issue involves replacement and hence retrenchment of workers through the introduction of technology. The clerical and sales areas are particularly affected and this is where most women are employed. In 1980 one public service union estimated that staff levels would be cut by up to 50 per cent on the introduction of word processors. A cut of 21.7 per cent over eighteen months was the result of automation in one major insurance union. A Mutual Insurance office introduced twelve word processors, with twenty-nine operators and two supervisors to replace 100 typists.[113] The statistics go on and on. These facts have spurred on government women's departments to encourage girls to seek occupations in other areas. They have been a prime motivating factor in encouraging girls to do science and maths in high school, which considerably widens their choice of university courses and therefore jobs.

Treatment of women workers by the trade union movement has always been a battleground for women. In September 1983, Jennie George became the first woman to be elected to the executive of the Australian Council of Trade Unions (ACTU). She stated then that she was particularly concerned about child care, wage discrimination and the number of women not covered by award provisions.[114]

The proportion of women in unions in Australia is lower than that for men. In December 1983, 43 per cent of waged women and 61 per cent of waged men were union members, though the increase over the last ten years has been greater for women.[115] In North America only a small proportion of the labour force is unionized—20 per cent. Women's participation rate is only 11 per cent. This is primarily because of the historical failure of the unions to cater for women's needs.[116]

A number of myths about women and unions have been drawn up to explain their lack of participation within them. One is the myth of the passive worker: that women cannot bond to strike and are never militant. History shows the fallacy of this, women have been active in the development of the labour movement and there have been women's strikes and militancy within women's unions.[117] Brady, for example, has shown that within the hotel and catering industry women have always worked to improve their conditions and their 'collective response shows a long term involvement in trade unionism'.[118] They were members, leaders and took an active part in the union. In unions where this is not the case, structures often operate to exclude women from action and leadership. Their family responsibilities mean that out-of-work-hours union labour is a triple burden, even if they could break the male stranglehold on leadership roles. In addition, incentive to participate may often be low when women are in male-dominated organizations which do not recognize that access to abortion for example is a work issue, and that child care is essential to women workers.

Some change may be coming, indicated by the success of Jennie George. But these changes are hard won, both for the individual and for women as a group. Impetus for unions to include women has come from pressure by women themselves, the anti-discrimination laws, declining membership and lessening power of the unions particularly in North America, and women's increased presence in the workforce.

Conclusion

Women's labour, has been devalued by men, who benefit from the negative self-definition this gives women. Self-respect is a prerequisite of action for change and waged labour in our society gives women that self-respect through increased status and power. Trends now indicate that women increasingly resist doing the housework and rearing numerous unplanned children.[119] Most are increasingly seeing their economic security, not in men who frequently abuse the power or abdicate their financial responsibilities to families, but in being self-supporting.

Barbara Ehrenreich has commented that 'it is almost as if the economic stresses of the 70s split women into two groups: those who went *out* to fight for some measure of economic security (either out of necessity or choice, though the distinction is not always a meaningful one), and those who stayed at home to hold on to what they had'.[120] Women in the workforce and women in the home are trying to gain control over their lives and the power to make decisions about those lives. But they are both working within a society which values only paid labour; which has a high unemployment rate; and which does not recognize women's unpaid emotional and domestic labour, let alone pay for it.

Power can be defined in terms of the access, or lack of access, a person has to resources (e.g. financial and educational), to status and to self-confidence. These are rarely found within the role of dependent houseworkers yet all accrue in some measure to working women. No matter what *ideal* society women would like to create, this is in *reality* the situation within which women live in Australia, and why they claim the right to paid employment.

5

Love, Sexuality and Friendship

Social relationships are part of the moulding of self. Intimate relationships have an even stronger impact on the confidence of a person, on a sense of who she is, and on her self-perceived value. Friends, lovers, family, reflect a person back to herself.

Love, sexuality and friendship form the core of intimate relationships. But these have been used and abused by men in the social control of women. Men have defined women's love, sexuality and friendship as valid only in relation to them, primarily within heterosexual marriage. Submission to men and fulfilment of their lives by proxy is part of the patriarchal definition of femininity.

Woman has been defined as being *for* man: woman should relate emotionally and sexually only to man. Janice Raymond created the concept of 'hetero-reality' to describe this relationship. She writes:

While I agree that we are living in a heterosexist society, I think the wider problem is that we live in a hetero-relational society where most of women's personal, social, political, professional and economic relations are defined by the ideology that woman is for man.[1]

In this kind of society 'woman-to-woman' relationships are seen as immature or unimportant. The mature woman is the woman who relates to man.

Part of the struggle for an authentic identity for women has entailed resistance to the imposition of heterosexual marriage and the determination to love women in friendship and sexually in order to create women- and self-identified selves.

Intimacy and the self

People need close relationships as a source of individual strength. They are necessary to make a person feel less alone, to feel known, to be loved and to love in return. Bardwick has commented that 'to be intimate is to reveal, need, hear and give. With a sense of mutual intimacy we have a feeling of mutual commitment.'[2] Intimate relationships are important because they foster self-confidence; someone in the world knows you well, and loves you.

The ability to form and maintain intimate relationships has been reduced by patriarchy to a purely sexual relationship. In general men fear intimacy and devalue it. One reason for this is that by making the self vulnerable in intimacy, a person must accept a certain level of powerlessness. Within intimacy lies also vulnerability, a certain dependence and trust.

Women are more successful at forming intimate relationships. In discussing the development of the sexes, Chodorow points out that through their differential development, girls emerge 'with a basis for "empathy" built into their primary definition of self in a way that boys do not. Girls emerge with a stronger basis for experiencing another's needs or feelings as one's own (or thinking that one is experiencing another's needs and feelings)'.[3] So girls experience themselves as more continuous with others, 'their experience of self contains more flexible or permeable ego boundaries'. Boys, on the other hand, define themselves as more separate and have rigid boundaries of self. So while a woman's sense of self is 'connected to the world', 'the basic masculine sense of self is separate'.[4] Men become inexpressive, 'emotionally constipated', and emotionally superficial, none of which are conducive to communion and attachment.[5] Empathy is an essential part of women's relationships with others and involves the anticipation of another's needs; the understanding of another person's experience as her own. Men 'find it difficult and threatening to meet women's emotional needs'.[6]

Included in intimacy is the ability to self-disclose, which women are more ready to do, opening channels of communication with others. But self-disclosure also gives the other person power by giving them potentially damaging or manipulable information over the self-disclosing person. By withholding information about themselves men place themselves in the more powerful position. Women's vulnerability is increased because they 'are socialized to *care* more than men—especially about personal relationships'.[7]

Male writers have attempted to analyse the problems men have in relating to people intimately. Vincent Kavalovski has pointed out the competition, deep distrust and lack of intimacy which marks men's relationships with other men.[8] Joseph Pleck points out that self-interest is the primary motivation for men. He explores the way they experience their emotional lives through women and their resentment of what they see as this form of female power over them. Men attribute to women a 'masculinity-validating' power: 'to experience one's self as masculine requires that women play their prescribed role of doing the things that make men feel masculine'.[9] Men are frightened that women will withhold these, thus burdening women with the responsibility for making men happy. Through the institution of marriage, men have tried to ensure that women relate sexually and emotionally primarily to them.

Men and marriage

The closest experience most women will have with men is within marriage, either de facto or de jure. Women are said to be eager to marry because they need to be loved by men, while men theoretically hold back more, because in marriage their 'freedom' ends.

It is a relatively modern assumption that love is a reason for marrying. Traditionally, the acquisition of property, an unpaid workmate, or children (particularly a male heir) were the primary reasons for men seeking marriage. It still remains an exchange, a bargain in terms of men providing shelter, financial support and protection; while women supply sexual accessibility, children, emotional and domestic support, as well as often financially contributing. Marriage is an exchange of service resembling, according to Rowbotham, the bond between man and man in feudal society: 'The woman essentially serves the man in exchange for care and protection, though the specific balance between them is personally determined'.[10]

Power within marriage has been shown to be determined primarily by the economic power of individuals outside the marriage. Wives who are in paid employment, though they face twice the workload of non-workforce women, experience greater satisfaction with marriage and have greater power within it.[11] Apart from the fact that paid work is more highly valued in our society, the 'employed' woman increases her bargaining position through the economic contribution she makes to the marriage and the added role she acquires through work. Walum has pointed out that 'in all human exchanges, economic or romantic, the person who has the least interest in the exchange has the greater power, because she or he can easily withdraw from or terminate the exchange'.[12] The increasing tendency for women to initiate divorce and separation has occurred as independent incomes become more accessible to them.

Marriage today is romanticized to include 'love'. But it is difficult (Firestone says impossible) within this exchange system to create a loving and intimate relationship, because women are economically dependent and a power imbalance operates. The love that men demand from women in marriage differs from the understanding women have of it. Men do not 'commit themselves to mutual openness and vulnerability until they are forced to'[13] so women have continuously to cloak their own selves in favour of the male vision of woman. They restrain their love and need for emotional commitment, hoping to adjust to the reality of marriage. Nancy Friday sums up the feelings of exploitation women feel when they think of love:

We are the loving sex; people count on us for comfort, nurturing, warmth. We hold the world together with the constant availability of our love when men would tear it with their needs for power. We feel incomplete alone, inadequate without a man, devalued outside marriage, defensive without children. We are raised for love, but when love comes to us, sweet as it is, somehow it is not as ultimately satisfying as we dreamed. We are being loved for being a part of a relationship, for our function—not for ourselves . . . Society plays us a dirty trick by calling us the loving sex. The flattery is meant to make us proud of our weakness, our inability to be independent, our imperative need to belong to someone.[14]

Firestone has severely criticized 'love' as being 'the pivot of woman's oppression today'. By this she means, not the open exchange of caring which love can be, but what love has become, defined by men: a total submerging and submission of woman's self to the physical and emotional service of man. She writes: 'for millennia they [women] have done the work, and suffered the costs of one-way emotional relationships the benefits of which went to men and to the work of men'. This love, often termed 'romantic love', is packaged and sold to women through, for example, advertising. So men have created a selfish version of love, not 'an exchange of selves', but a relationship in which the male 'self attempts to enrich itself through the absorption of another being'.[15] In love then women are expected to self-destruct, to de-self. As Friedan wrote, love, for women, has been defined 'as a complete merging of egos and a loss of separateness—"togetherness", a giving up of individuality rather than a strengthening of it'.[16] The danger in accepting male definitions of women's love is that in order to reduce their feelings of alienation, women may indeed submerge themselves in family, identifying with others 'to the extent that self is defined externally, through others—she becomes nothing but them'.[17]

Lazarre confronts the difficulty of a long-term marriage which is not a 'bad' marriage in the traditional sense, but one in which the self is gradually and subtly eroded. Women may break from a seemingly comfortable relationship with a drive toward renewed self-consciousness. Speaking of some of these women, Lazarre analyses their constant renewing and writing out of their past relationship in an attempt to understand this seemingly unnecessary break: 'Women alone seek consciousness or solace. Women married often lull their spirits into a death-like fever'.[18]

Male authority is institutionalized within marriage. It is important to men to control the institution of the family because 'where families are basic units of social organisation, there are political advantages to be had in being their managing directors'. Oakley has argued that men's right to control their wives in whatever way they choose is 'another item on the covert agenda of sexual relations behind the tempting fantasy of a romantic idyll'.[19]

The ideology of privacy has assisted them in defining family behaviour as outside the 'interference' of the state. The assumption has been that men will 'protect' members of their family. But social control is intricately intermeshed with personal experience. Marriage is 'not isolatable or separable from the other social institutions' which men have constructed and control.[20] Women are encouraged to believe that happiness will be found within marriage but many discover that it is marriage that oppresses, and depresses, them. Married men are mentally healthier than single men, but married women are less mentally healthy than single women and married men.[21] The self-esteem of married women is generally low. Unger points out that the disparity between the reality of marriage and the expectations of it is a shock, particularly to women with high self-esteem.[22] Marriage is good for men but not so good for women.

The physical vulnerability of wives is institutionalized within marriage and there is little redress for women against assault or marital rape.[23] Wife-battering is becoming more widely acknowledged, though the *reported* incidence hides the many unreported assaults. 'Like other crimes which take place within the family, domestic violence is shrouded with shame, fear and secrecy.'[24] It is impossible to assess the number of women who, having been assaulted once or more, live either in terror or in a situation where they take care constantly lest the assault recur.

Seeking the 'protection' of one man against the many fails. It fails in respect to women themselves, who may be physically abused and raped within marriage; it fails in respect to economic security, as women have no *legal* access to their husband's salary and rely on goodwill; it fails in respect to the protection and support of the children of the marriage, as evidenced by the rates of father—daughter incest and the refusal of 75 per cent of divorced Australian men without custody of their children to pay for the maintenance of those children after divorce.[25]

Male-defined sexuality

Within marriage sexual intercourse is seen as part of the bargain women agreed to, and there is no right to refuse. Male access to the woman's body is undisputed in law and within the ideology of marriage. So the woman *herself* in addition to her labour has been exchanged for shelter and protection. In the minds of many women, the two become synonymous: labour includes sexual intercourse.

For some women, then, marriage is a form of prostitution. Lindsey has commented that the structures of sexism and economic oppression which men have constructed ensure that 'to survive within those structures, all of us, all the time, in some way or another, sell ourselves to men'. She argues that by setting 'moral' women against prostitutes as a group, men divide women, blinding them to the reality of their united condition. 'We are taught that prostitution is a deviation from the social—sexual norm. In fact, it is the most blatant manifestation of the norm itself'.[26]

Heterosexuality is the foundation of this oppressive norm. As Charlotte Bunch wrote: 'heterosexuality means men first. That's what it's all about. It assumes that every woman is heterosexual; that every woman is identified by and is the property of men'.[27]

Adrienne Rich took this up in her analysis of 'compulsory hetero-sexuality' as a political institution: a means through which men have alienated women from each other. She discusses the use of pornography to objectify women and to present sexuality and violence as congruent, representing the myth that 'for women sex is essentially masochistic, humiliation pleasurable, physical abuse erotic'. Pornography not only creates the equation between violence and sexuality but 'it widens the range of behaviour considered acceptable from men in heterosexual intercourse'.[28] Rich cites the ideology of heterosexual romance as the tool used by 'the procurer' to 'befriend' the victim. So the prostitute is conned into seeing her pimp as a friend, and 'love' bonds the incestuous father and brother to their victims and convinces the battered wife to accept her husband.

Dixson comments on the immaturity and ugliness of the Australian male's view of sexuality. Women are still viewed as dirty and sexual intercourse called a 'naughty'. She quotes Craig McGregor who writes that men

regard women merely as sexual providers, things to sleep with but not to talk to, and the extraordinary prevalence of phrases such as 'they're all the same in the dark' and 'makes no difference with a bag over her head' illustrates an attitude to women which is narrow, cynical and immature.[29]

Dale Spender points out that language defines sexuality from a male perspective. Women's sexuality has been defined in 'minus' terms—lacking a penis. There are words to describe a sexually healthy male—'virile', 'potent'—but only negative terms to define sexually active females—'nymphomaniacs'.[30] The sexuality of childbirth and lactation is completely ignored. Because men cannot see anything sexual in birth they will not allow women to see it as a potentially erotic experience. Sexuality is thus defined as belonging only to the dominant group.

The 1960s wave of the women's movement placed stress on the validity of the clitoral orgasm and independent sexuality for women. It attacked marriage (sex) manuals which extorted the husband to be assertive and masculine, leading the way with his greater experience, while the woman was to satisfy in her passive way. If he failed to bring her to orgasm, she was labelled 'frigid' and a failure.

There is some debate over who first 'discovered' the clitoris, as Campbell points out, commenting that 'presumably its discovery was as much a surprise to its owners as the discovery of America was to the indigenous occupants'.[31] Laws and Schwartz comment on this as evidence of the power of sexual scripts as social institutions regardless of women's *experiences*:

It is evidence of the power of social constructions over subjective reality that the experience of women was set aside, decade after decade, in favour of the authoritative pronouncements of Freud, who was of course physically incapable of experiencing a vaginal or a clitoral orgasm much less discriminating one from the other or judging one as superior to the other on the basis of his own subjective experience.[32]

A number of research studies by Kinsey (1953), Ellis (1953) and Masters and Johnson (1966) had moved toward 'revalidating' clitoral orgasm and debunking the Freudian myth that only the mature, vaginal orgasm defined women's pleasure.[33] But sexuality was still defined as the heterosexual act, where penetration was supposed to stimulate the clitoris. Yet Hite has reported that only 38 per cent of 2000 women surveyed ever experienced orgasm in intercourse.[34]
Like many issues that new wave feminism dealt with, women had debated sexuality before. It has been a constant in women's battle for self-empowerment. Elizabeth Wollstenholme Elmy and Frances Swiney, both suffrage fighters in England in the late nineteenth and early twentieth centuries, wrote extensively on sexuality. Wollstenholme Elmy was indignant of the sexual coercion allowed within marriage, and campaigned to have women's rights to refuse sex within marriage enshrined within the law. Other issues taken up actively were the fight to protect children from sexual exploitation, and refuges to help the victims of rape and abuse; the struggle to have women police work with the victims; to have women doctors treat the cases; and have women-only carriages on trains. The extensive 'white slave' traffic was fought.
Feminists argued continually that the way to stop venereal disease was by male abstinence, not the repressive laws enforcing medical checks on women.[35] The alarming effects of venereal disease upon women were angering feminists, and in 1913 Christabel Pankhurst wrote *The Great Scourge and How To End It*, claiming that 75 to 80 per cent of men had gonorrhea and a considerable number had syphilis. She detailed the effects on wives and children. Her catch-cry was 'votes for women, chastity for men'. In Australia, by 1927 the Victorian Education Department claimed that 7.5 per cent of the total Australian population was affected by venereal disease.[36]
The issue of 'free love' which emerged at the end of the Victorian period also divided feminists. Vida Goldstein publicly declared that she was *not* in favour of it as this was a claim used to discredit feminists. Though often labelled prudes and puritans, the anti-sex feminists saw the new emphasis on free love as an imposition of male sexuality on women: it only made women 'free' *for* men. Others feared the spread of disease. But some feminists saw it as a freedom for women in breaking with the constraints placed on their own sexuality.
Entwined in these debates were those on abortion, birth control and sexuality with children. In all these discussions in the nineteenth century, women can be seen 'manoeuvring, both to gain more power and control within their own lives, and to remove the restrictions placed upon them by the exercise of male sexuality inside and outside the home'.[37]

The break between reproduction and sex became a reality for women in the 1960s when the contraceptive pill became available. The issue of control now seemed solved, women were free to indulge themselves sexually because they could control the possibility of pregnancy, However, pressure was now brought to bear on them because of this freedom: women should do it all the time with anyone who asked. Women's sexuality was still defined in relation to men but now it resembled that of men more closely: the one-night stand, no-care-needed quickie. (The contraceptive pill itself brought a new variety of health problems for women, including thrombosis and cancer.)

The difficulty for heterosexual women lies in defining their sexuality outside male definitions. But it is often impossible to ignore the continuum on which male sexuality itself resides, and its relationships to power. Andrea Dworkin has written that, alongside the persuasive ideology of 'romantic' love lie rape, domestic violence, pornography, prostitution, and brutal practices such as clitoridectomy.

Dworkin stresses the issue of power in sexual relations between men and women, stating that women have had to come to terms with forced sex as part of normal life. 'Forced sex, usually intercourse, is a central issue in any woman's life. She must like it or control it or manipulate it or resist it, or avoid it; she must develop a relationship to it'.[38] She most definitely cannot ignore it.

Beatrice Campbell discusses a study by Dickinson and Beam (1932) which was the first major medical analysis of marriage and revealed 'dreadful pain and maladjustment' in women. Of the 1000 cases that Dickinson and Beam studied only 363 were 'adusted' sexually, that is, 'without complaint'. Suggestions were made that in this study frigidity could be seen as 'covert rebellion'. A different survey of about 150 women indicated that the pain of sex for some women was not frigidity at all, but 'lack of consent', raising the unspoken issue of forced sexual intercourse within marriage.[39]

Women have had to contend with sexuality supposedly sealing the intimacy of a heterosexual relationship while at the same time watching it used as a tool of male brutality and power in rape and incest. Many writers discuss the fine dividing line between 'normal' and 'abnormal' sexuality, leading to the conclusion that there often is little difference between sexuality within marriage, within prostitution and within rape.

Catharine MacKinnon writes: 'as women's experience blurs the lines between deviance and normalcy, it obliterates the distinction between abuses *of* women and the social definition of what a woman is. That is, to be a woman is to be sexually used/abused; to exist for someone else, who is male. Women's sexuality becomes defined as 'the capacity to arouse desire in that someone'.[40] In this situation, woman does not have a unique and valid sexual identity of her own. She is the passive receptor of male sexuality; the object to which 'it' is 'done'. Language reinforces the male definition of sexuality: women are 'laid', 'screwed', 'penetrated'; they can be a whore, slut, or nymphomaniac but never 'potent' or 'virile'. There is, in fact, no term for women's normal sexual power.

Coveney and her colleagues problematize male 'normal' sexuality, delineating its characteristics: power, and the need to dominate, which they see as reflected in 'normal' sexual behaviour; the expectation that *he* will initiate sex; the selection of a smaller and younger partner; and the preference for sexual positions in which the man is in control.

Aggressive male sexuality is part of the male arsenal in anti-woman society, exemplified in statements such as 'what she needs is a good fuck' directed towards 'uppity' women as punishment. Rape fantasies and pornography are further examples, and the extreme case is the rape/murder of women.

Coveney et al. include the separation of sex from loving relationships as part of male sexuality, an attitude encouraged in women during the 'free love' period of the 1960s and 1970s. Importantly, 'this requires that men develop the ability to objectify women'.[41] Men treat women as objects in advertising and pornography, but also through, for example, whistling or lewd comments. They make a fetish of parts of women's bodies—'I'm a legs man'—separating them from the whole woman. And finally, male sexuality is characterized by its supposed uncontrollability: the rationale for rape.

Within this context it is difficult for women to develop reciprocal relationships with men which acknowledge the links of sexuality with intimacy, the difference in female and male sexuality, and the elements of power in male sexuality which need to be eliminated. The development of authentic loving and sexual relationships with men is difficult within the differential power relationships between the two sexes. But women continue to try.

In a moving commentary on 'loving men', Jane Lazarre stresses the need to explore women's relationships with individual men which are positive and loving. She comments on the difficulty of discussing an issue which is 'so central' to many women's lives, an experience which is intense, diverse and confusing.[42]

Passionate sexual love is one way of breaking the 'protective fog in which we lose the sound of our own voice'.[43] The difficulty even with sexual passion, however, is in seeking fusion while not losing the self. Women must demand and ensure autonomy of self and fight engulfment and oppression within the politics of intimacy in female/male relationships.

Marriage resistance: single women

The pressure on women to marry is enormous within patriarchy, a woman's identity and status being measured by how successful a 'catch' she makes. Single women are seen as a threat to the social order because they have stepped outside the 'caring and bearing' system.

Hetero-reality means that women cannot be seen to exist except in relation to men. Women today who dine on their own will witness this, when they are often not served in a restaurant or even standing at a shop counter. A woman alone becomes invisible. Janice Raymond cited Lily Tomlin's version of this experience: 'I've actually seen a man walk up to four women sitting in a bar and say: "Hey, what are you doing here sitting all alone?"'. Single women have been defined as lewd or loose, indicating promiscuity. But 'loose' basically means unattached. As Raymond writes:

For men, any free and independent woman signals a 'loose woman'. The freedom of women is equated with whoredom. A woman 'on her own' is regarded as 'free for men'. Looseness is reduced to lewdness.[44]

So the state of being free and unattached to any man becomes the negative state of being available to any man.

Margaret Atwood's chilling novel *The Handmaid's Tale* (1986) also uses this concept when she discusses the escape of Moira from the school where women are being trained to be surrogate mothers after a right-wing revolution. Atwood captures the expansiveness and freedom of being 'loose':

The thought of what she would do expanded until it filled up the room. At any moment there might be a shattering explosion, the glass of the windows would fall inwards, the doors would swing open ... Moira had power now, she'd been set loose, she'd set herself loose. She was now a loose woman.[45]

Historically, spinsterhood has often been deliberately chosen as a lifestyle, causing concern among men. For example, in the middle of the nineteenth century in England there was an extraordinary public debate over women's capacities for friendship and communal activity. Part of the concern was generated by a preoccupation with a strong demographic imbalance in the population. This was partly caused by the exodus of men to the colonies. In the census of 1851 there was an excess of women in the population of more than half a million. In 1862, Frances Power Cobbe pointed out that there was 'an actual ratio of thirty per cent of women now in England who never marry'.[46] Many male writers were bemoaning this terrible fact, asking whatever would the single woman do without a man? But women were obviously not single solely because there were no men to marry! In fact as Cobbe pointed out there was a considerable percentage of old bachelors whose plight excited no interest at all. The belief was that no *woman* could possibly be single by choice, this being an 'unnatural condition'.

The fact remained that one in three of all adult women were single and one in four would never marry in the Victorian time. Seeing these women as superfluous, many men suggested that they be forcibly emigrated. But they did see that there was a resource problem involved in this as it would require about 10 000 ships or 10 000 voyages across

the Atlantic and to Australia in order to accommodate the surplus![47]
Many women have made political decisions to be single. Christabel
Pankhurst stated categorically that spinsterhood and celibacy should
be a choice made in response to the conditions of what she called
sex slavery. She stated: 'There can be no mating between the spiritually
developed women of this new day and men who in thought and
conduct with regard to sex matters are their inferiors'. In the Women's
Social and Political Union, with which Pankhurst was involved, 63
per cent of members in 1913 were spinsters and many of the rest
were widows.[48]

Women rebelled against the sexual subjection of women. Frances
Swiney in 1907 accused men of having reduced women to a purely
sexual function. She wrote: 'Men have sought only a body. They have
possessed that body. They have made it the refuse heap of sexual
pathology . . .'.[49] Swiney's solution was that sexual intercourse should
take place only for the purposes of reproduction.

Other women such as Annie Besant, originally a practitioner of free
love, changed their opinions to become energetic proponents of celibacy
as a lifestyle. In fact a whole section of the women's movement built
up before World War I advocated remaining celibate and unmarried.
Many of these women were involved in the social purity campaigns.

Cecily Hamilton wrote a book called *Marriage as a Trade* (1901) in
which she saw the enrolment of spinsters to the women's revolution as
a political necessity. If spinsters showed their lives to be full and
happy, they would produce for other women a viable alternative to
marriage and sexual degradation.[50]

This growing group of women met with serious opposition. From
the 1890s the profession of sexology, instigated by Havelock Ellis and
his comrades was developing. Sex reform was gaining popularity. It
was argued that heterosexual sex was a physical necessity for both
sexes and that repression was very dangerous.

Single women today are stereotyped on the one hand as miserable,
abnormal failures or, on the other hand, as irresponsible, sexually
'loose' and uncaring. They are attributed a negative image in order to
coerce women into the married state. 'Their rejection of traditional
male protection sets single women up as ready targets for male hostility'.[51] Single women are seen as a threat to the established order
because they proclaim the possibility of being fulfilled without living
with a man.

Studies indicate that in fact single women are on the whole happier,
more mobile and less anxious and depressed than their married sisters.
They tend to be middle-class, educated and in good occupations.
Educational and financial status may influence the choice to remain
single. Unger writes that if a working-class woman chooses singlehood
she is more likely to experience economic difficulties and is often
trapped within her family situation.[52]

Women who say they *choose* to remain single are often simply not believed, and are ridiculed or pitied. Psychological theories support these assumptions and have regarded singleness as an immature and even deviant state. Marriage is defined as normal for women. The disadvantages in being a single woman stem mainly from social pressure rather than the state of singleness itself. In their study of Australian single women, Penman and Stolk note that living without men has its problems: the lack of a male 'protector'; the emphasis on couples in social structures; and possible loneliness.[53] The argument that a woman cannot live safely without a male protector is fallacious but the very fact that a woman has a relationship to a man serves to give her the *feeling* of possible protection.

Though there are disadvantages to being a single woman in a couple-oriented society, there are many advantages. Privacy and solitude provide an individual with time and space in which to develop a deeper self-knowledge, a sense of separateness and wholeness, and of self-sufficiency.

Singles can work evenings if they choose, change jobs, travel and organize their lives solely around their own needs. They have the freedom to choose a sexual partner, solitude, company, work, a timetable. Work can become a central intellectual and social pivot in their lives. Donelson and Gullahorn point out that a cause of some kind may be 'an integrating core of life'. It may be 'so viable, vibrant and alive that a commitment to it is a relationship with a living organism'.[54]

Developing intimate relationships is difficult and often hard work. Penman and Stolk note that the ability to form intimate relationships may well be an indicator of psychological maturity, but 'marrying does not facilitate or demonstrate this ability'.[55] Living alone can indicate the struggle for a life which has less compromise and gives a woman greater control over herself. With an echo of Raymond's comments about loving the Self, Bickerton comments that: 'After all, if it is not possible to be alone, then it is usually not possible to be close'.[56]

For single women, securing a constant and satisfying sexual life involves a number of difficulties: decisions about casual sexuality, relationship sexuality, abstinence or celibacy, heterosexuality and homosexuality. The risks to a woman in sexual liaisons include possible violence, exploitation and communicable diseases. Many find the risk in casual sexual liaisons worthwhile because they can retain freedom and control; but many feel exploited or feel that they are exploiting the woman/man involved. The risk in longer-term sexual relationships is the loss of independence as demands of the relationship grow, though for many single women the attainment of a satisfying relationship is still a sought-after goal.

Of the 435 single women interviewed in the Penman and Stolk survey, one-third had never had sexual intercourse. Half of those who had not experienced sex did not want to. Penman and Stolk comment that 'in over-emphasising sexuality we have devalued by neglect, the role of celibacy'.[57] There are, contrary to popular myths, no grounds for assuming negative effects from celibacy. Many women, feminists and non-feminists, now advocate celibacy as a lifestyle ensuring self-respect. Dale Spender has commented that the way sexuality has been defined by patriarchy 'repels me' and Connie Purdue wonders 'How can men to whom sex activity means so much ever understand the ease with which women can lead a satisfactory life of sexual abstinence?'.[58]

Kwitney discusses the advantages of celibacy after a divorce and a number of casual sexual relationships which became increasingly unsatisfactory. Likening the feelings to those of a fast, in which she experienced a 'denial of my hunger in service to myself', Kwitney comments on her increased sensitivity to her own body, a heightened awareness of herself, an enjoyment of the low-level sexual tension which has a 'vitality and excitement of its own'. The longer she was celibate the more centred she felt herself to be. She found it a very sensual period, yet one of peace with herself, with a sense of completeness. She writes: 'I see that there is some work on the self that can only be done alone, independent of relationships. That work is the affirmation of one's self'.[59]

Motherhood

The other loving allowed to women under patriarchy is mother-love; again a self-denying, self-sacrificing love. The institution of motherhood as men have defined it includes the economic dependence of women and their enslavement in child-rearing. It also ties them into heterosexuality and marriage.[60] Malestream dominant culture has worked to divorce women from the experience of motherhood as women would choose it, because of the fear men have of the procreative power of women.[61]

Adrienne Rich has explored the institution of motherhood as it has been controlled by men to the point where it is a distorted experience for most women. It is this *institutionalization* of motherhood which is the problem, not the experience itself, or the potential of woman-defined mothering.[62]

Motherhood under patriarchy causes many women to experience depression when they face an unexpected decrease in power simply because of child-bearing.[63] Their physical dependence is often followed by economic dependence for themselves and the child. Furthermore, when they have children women invariably have less time for servicing their husbands' emotional lives, which men resent.

Though the ideology of women's fulfilment through motherhood explodes through the media and other patriarchal institutions of ideological reinforcement, little practical reward or support follows. The intention is to tie women into sole responsibility for children, witnessed, for example, in the difficulty in obtaining government-funded child care. The mythology of the social importance of motherhood is at variance with women's experience of it.

Elizabeth Badinter analyses the development of maternal instinct as a construct, showing that 'no universal and absolute conduct on the part of the mother has emerged'.[64] She concludes that the difference between giving birth and having the sole responsibility for rearing the child have been conveniently confused by patriarchy and that is the key to women's oppression. Chodorow traces the way mothering is passed on from mother to daughter. She writes that by and large women want to mother, and get gratification from their mothering; and finally, that, with all the conflicts and contradictions, women have succeeded at mothering.[65]

Barbara Wishart, an Australian lesbian mother who conceived her daughter through artificial insemination by an unknown donor, writes that the 'experience of motherhood has given me a deep bond with other mothers I know, and a sense of continuity not only with the women in my own family, but also with the continuous line of women from antiquity to the present day who have borne children'. She stresses, as does Rich, that it is the structure within which motherhood is experienced which oppresses women and we need not devalue, as patriarchy does, the experience of motherhood itself which still has 'something *positive* or *worthwhile* or even *wonderful* about it'.[66] Motherhood has become a perverted experience under patriarchy, often entailing hatred, anger and resentment, particularly between mothers and daughters. Women de-selfed through mothering become anti-nurturing. As Sara Ruddick writes:

The more competitive and hierarchical the society, the more thwarted a mother's individual autonomous pursuits, the more likely that preservation will become egocentric, frantic, and cruel. Mothers recognise these dangers and fight them.[67]

Rich points out that the nurture of daughters within a patriarchal society can only be achieved by mothers with a 'strong sense of *self-nurture*'.[68] Yet this is constantly eroded and undermined by patriarchy, which enslaves women within the institution of motherhood, idealizing it while failing to support it; using and exploiting women's nurturing skills while derogating them. The relationship of mother to daughter is seen as threatening to men. They react by neglecting it and by over-stressing the mother—son relationship; or they trivialize and try to break it, as exemplified in anti-mother-in-law jokes which men continuously perpetuate.

The sexual politics of motherhood are learned from our mothers: women either grow from, imitate, or react to and reject, the mothering they themselves receive. Emotional ties remain in the relationship, whether they be positive or negative. Because patriarchal control of women's lives and thus identity has cut off a sense of their self-worth from women, mothers often pass on this heritage to their daughters.

This is done both from bitterness on the part of mothers who envy any move their daughters make to free themselves; and from love and concern that daughters learn to adapt and survive within patriarchy as they themselves have done. It is easier for daughters to reject their mothers than to accept the forces which act upon them and which in some ways exonerate them. Rich writes that daughters 'see their mothers as having taught a compromise of self-hatred they are struggling to win free of, the one through whom the restrictions and degradations of female existence were perforce transmitted'.[69]

The history of motherhood and the family from women's viewpoint has been a struggle to survive and to gain the best possible choices for their children. The questions of emancipation and equality have been related to motherhood, as children restrict a woman's mobility and economic freedom, tying her to the father of the children with few escape routes should she desire them.

The effect of patriarchal definitions of women's experience of mothering is often devastating. Women continue to experience it within a context of powerlessness. As Ruddick writes:

Throughout history, most women have mothered in conditions of military and social violence and often of extreme poverty. They have been governed by men, and increasingly by managers and experts of both sexes, whose policies mothers neither shape nor control. Out of maternal powerlessness, in response to a society whose values it does not determine, maternal thinking has often and largely opted for inauthenticity and the 'good' of others.[70]

Many women who in the past married in order to have children are now able to be single and mothers, though this is more likely for the economically independent woman. Artificial insemination by donor (AID) clinics report a number of single women obtaining pregnancies, and self-insemination groups for single women are operating particularly in Britain and North America.[71] Single women who are mothers by choice report it as a positive experience, particularly because they do not have the stresses and tensions of marriage and the emotional and practical servicing of a husband to contend with.[72]

Some women are infertile and cannot biologically mother. The power of the ideology of woman defined as mother makes their experience painful and alienating. Few women are born infertile. Causes of infertility include: IUDs, sexually transmitted diseases, post-operative infection and other iatrogenic (doctor-induced) causes. But women are made to feel blameworthy and guilty about their infertility, particularly if they have had more than one sexual partner or an abortion. A woman may feel particularly frustrated and resentful if she has been using, for example, the pill or an IUD for years only to find it was unnecessary. Mazor has described the testing process as 'assaultive', as women are required to 'expose their bodies for tests and procedures', and 'to expose the intimate details of their sexual lives and their motivations for pregnancy'.[73] Often the male partner was infertile, yet the woman has been subjected to lengthy, unpleasant and intrusive tests, the assumption being that woman is naturally 'the problem'.[74] In fact, men and women are equally represented in the statistics on infertile couples.

The force of social and self-disapproval for not mothering can lead infertile women onto reproductive technology programmes, attempting to 'try everything' including dangerous and unsuccessful technologies (see Chapter 6 for a fuller discussion).

But some women overtly resist the definition of woman as mother, choosing instead a child-free lifestyle. These women are labelled negatively.[75] They are seen as unlikeable, selfish and uncaring child-haters.[76] In one study of community attitudes in the US, researchers asked their respondents to judge the personality and character of both child-free couples and couples with children. People perceived the sterilized child-free wife as 'less sensitive and loving, less typical an American woman, more likely to be active in women's liberation, and less happy, less well-adjusted, less likely to get along with her parents, and less likely to be happy and satisfied at age 65' than an mother of two described as alike in other respects.[77]

The advantages which women see in remaining child-free are: they will not experience the career and job disadvantage of time out of the workforce; they can continue a freer lifestyle which involves travel and job mobility; and they will not risk interference with and erosion of the intimacy of their marriage relationship. These women also express a strong desire for self-development. One 73-year-old child-free woman wrote in 1982:

If you want to know what it is like to grow older without children I will tell you, it is marvellous. My neighbour, older than I and a friend younger than I, all without children, time and time again say to one another, 'thank goodness I never had children'. Never once has any of us ever regretted it, and many are the times we have rejoiced over it . . . We were the smart ones.[78]

For these women, choice within male-defined motherhood was no choice, so they exercised their own will and resisted the role of mother. For them, as for many others, it became 'the Ultimate Liberation'.[79] The cost in social disapproval is high however, and child-free women are constantly resented, slighted and socially isolated from families with children, on the assumption that they dislike children and in some cases as a form of punishment for their choice. As Rich has written: 'mothering and nonmothering have been such charged concepts for us, precisely because *whichever we did has been turned against us*'.[80]

Father-right

As women struggle and succeed at establishing viable and independent households which exclude men, men fight to take control back with an increasing push for 'father's rights'. Women have wanted men to be more involved in child care so they would be more *responsible* for their children. But in outlining legal changes foreshadowed in Britain, Sutton and Friedman conclude that 'what has resulted is a minimal change in caring and a significant move by men to increase their *rights* and hence, control'. They write that women in the 150 refuges for battered women in England will attest that 'when a man has access to a child, he is able to use the child to further his own interests and to control that child's mother'.[81] Increased father-right, instead of increasing paternal responsibility for children re-defines women and children as belonging to men.

This is not solely within marriage, as recent legislation in many countries including Britain, New Zealand, Switzerland, France, West Germany and all Australian states except Victoria has extended the rights of husbands and fathers to men for children born *outside* marriage. In a time when family structures are changing, 'the family' is said to be in need of support, but it is actually the position of men within the family which is being shored up.[82] The erosion of responsibility in men both for the women they supposedly support and for their own children is more and more in evidence. Even anti-feminist women loudly proclaim women's fear of the 'rebellion of men' and their distrust of men.[83]

In Australia in 1983 it was suggested that the government might force men to pay the maintenance for their children which the courts have determined. The responses printed indicate the failure of men to accept responsibility for their children. Glenn Martin, proudly pro-claiming himself as a Justice of the Peace, wrote to the *National Times*:

I think mothers have to be quite clear about what they are expecting . . . If they choose to leave a man then they have to become responsible for their own lives and for those of their children. If they want maintenance they should realize that it is humanly impossible for any such money to be free of other links with the man.

If the man leaves, it's no different. If you want the man to be responsible, you have got to see that in terms of personal involvement with the children, and not just as maintenance. And if the man doesn't want to be involved you have to consider that maybe you're better off anyway, even without his money.[84]

More recently, the campaigns by men to control women through maintenance have gained strength. At the end of 1986, the Australian federal government released its proposals to collect maintenance at source, that is directly from the father (or supporting parent) through his/her pay cheque. Statistics indicated that only one in ten single mothers received maintenance. Patrick Heffernan, from Parents Without Rights, a primarily male organization, said that fathers should only pay if they received the reward of the services of a wife and access to their children.

The group he represents are mainly men who 'have been denied access to *their* children by the courts, or by wives on the run from them'.[85] Mothers obviously should not work. Heffernan claims that women who have custody 'should not require a car' as they should be near public transport and the school. He is concerned that 85 per cent of these women have a boyfriend with a car, who has 'taken on the pleasures of a relationship, and stepped in to fill the former husband's role' so he should also 'take on the responsibility for that woman and *her* [emphasis added] children'.[86] In other words, women should be punished for custody, while men seek to control them through their children.

In her extensive eight-year analysis of custody battles in sixty-five countries, Phyllis Chesler shows clearly that maternal right has never been legitimized. She documents the brutality of these battles and the way mothers are rarely judged 'good enough' to have custody. The more traditional a woman has been—a 'good' mother in patriarchal terms—the weaker her grounds for custody. The intention of the law is to reinforce father right.[87]

The question of the need to increase the 'rights' of fathers are taking place within a context of the inequality of women and men: a context in which men have more power solely by virtue of their sex. Now in the name of 'equality' men are actually being given *more power* over their children. As Pollock and Sutton conclude:

From the encouragement of man's presence at childbirth to the promotion of the ideology of fatherhood to judicial decisions in the custody of children—men's rights and status are being extended at the expense of women and children.[88]

This increasing presence of men in control within traditional women's spheres can also be seen in the response to struggles by the women's health movement to improve the alienated conditions of birth which male-controlled medicine constructed after the demise of midwives. The breaking of the isolation for women now in maternity hospitals is represented by the presence of the father, not by a reassembling of the women's presence which used to be associated with the ritual.

Women have sought the increased presence of men, in the hope that the status and power of mothers would increase, tasks done by men being accredited greater value: but the increased role of fathers in child care has not meant a reduction in women's work. 'They do not come into childbearing as workers, they come in as managers whose decision-making power is derived from the structures of sexual inequality'. They *choose* their commitment and mothers are left with the rest. Men *play* with children rather than *work* at the more mundane jobs.[89]

Again, men gain power, but women retain principal identity and low status as mothers. Though herself uncritical of this, Elisabeth Badinter expresses the underlying anxiety beginning to be experienced by women when she writes:

Under the pressure exerted by women, the new father mothers equally and in the traditional mother's image. *He creeps in like another mother, between the mother and the child*, who experiences almost indiscriminately as intimate a contact with the father as the mother.[90] (emphasis added)

There is increasing suspicion that if men become further involved they will take over and eventually control the sphere in which women exercise *some* power. As the evidence on increasing father-right indicates, while women want to share, men want to control. The concept of sharing rights and responsibilities can only be relevant in a situation of equal power. But as Pollock and Sutton note, the idea of fathers' presence at childbirth and their involvement in child care is based on an apolitical perspective of family structure. The rhetoric of shared parenthood glosses over these political relations and obscures the effects upon women and children.[91]

So as men combat women's growing independence from them, they are attempting to reassert control through children and increasing their legal 'rights'. But there is also an insidious reinforcement of negative self-definition for girls and women taking place within the family—through incest.

Ward points out that terms like 'incest' or 'molestation' protected people from recognizing the facts of father−daughter rape. The terms 'imply that something *less* than rape occurred' yet women of all ages describe the experience as 'feeling like rape'.[92] The statistics for Canada, Australia and the US indicate that 80 per cent of sex crimes committed against children occur within affinity systems (close relatives or friends); 50 per cent of these offenders are fathers; 5 to 15 per cent of the population are victims; and during wartime, incest declines. In line with other figures on child abuse, 92 per cent of incest victims are female and 97 per cent of offenders are male.[93]

In the USA the Children's Division of the American Humane Association estimates a minimum of 80 000 to 100 000 children are sexually 'molested' each year. Statistics are only now being regularly compiled and most are reported cases or those discovered by social welfare services—that is, figures are based on those families *not able to hide it.*

Rape by a father often continues for a long period of time. His access, after all, is virtually unlimited. In a study of fifteen women in therapy in the US, Herman and Hirschman report that the incest in these cases occurred when the victim was between four and fourteen years of age, repeatedly, and it lasted three years or more. In all but two of these cases it remained a secret.[94] Ward reports that the median age of victims is around ten years old, though there are examples of victims at two months, seventeen months and three years of age.[95]

Daughters who try to report incest are often accused of lying or fantasizing, even by professionals. The child may be exhibiting 'problem behaviour' because of the incest, and so invalidates her claims in the eyes of professional 'helpers'. The girl child often feels guilty; somehow responsible for her father's actions. In line with the myths of woman as the evil seducer discussed in Chapter 1, reasons are sought as to why the girl child *seduced* the father. Many 'helpers' are male, unwilling to accept the daughter as a rape victim, unwilling to accept their 'brother' as a child abuser.

The results of incest make relationships as adults with both women and men difficult for the victim. Some feel a sense of isolation from others, created as a protective defence. Faith in security and protection is lost. In their survey of 340 incest victims, the Sydney Women Against Incest group found that the greatest problem areas were related to family relationships, relationships in general, sexuality and mothering. Many respondents reported a negative effect on the quality of their relationships with their children, including a constant anxiety about protecting them, a fear of their sexual abuse and a concern with expressions of the child's sexuality.

The general effect on relationships was the greatest problem, involving distrust of both women and men. The women reported feeling alienated, isolated and different from other women. They had difficulty making friends because of the feeling that intimacy and care go hand in hand with abuse. The sense of the betrayal of trust could not be removed. Relationships with men were in general very negative, and sexuality was generally viewed unfavourably.

Almost all the women reported body-image problems, such as anorexia, being overweight, and bulimia. These were related to a compulsive cleanliness and feelings of ugliness. Many attributed those to the incest, reacting against the body which attracted the father.[96]

Although some mothers do act in this situation to protect their daughters, the privacy of family life makes it difficult to ascertain how they do so and the results. Most mothers find it difficult to act. Sylvia Kinder remembers her mother taking action on her daughter's rape; her father being sent by the court for treatment; her mother accepting his 'cure' and refusing to see that he continued to abuse his daughters.[97] Of the fifteen women discussed by Herman and Hirschman, only two told their mothers: one had her daughter committed to an institution; the other took the father to court, but when she 'realized that a conviction would lead to his imprisonment, she reversed her testimony and publicly called her twelve-year-old daughter a "notorious liar and slut"'.[98] Mothers transmit the message to their daughters: father first; it's dangerous to fight back; lose him and I lose everything; I cannot defend or protect you. Still relying on their abusing father for love and protection, the girl cannot walk away or hate him.

The mother is a primary target of blame in incest. She is blamed for collusion in the rape; for not 'seeing or hearing'. But as Ward points out:

No woman, except one who was a girl-child rape victim herself, has *any* information on which to base belief in such an event. All the cultural baggage about marriage, motherhood and Happy Families contains absolutely no information about the possible need to protect our daughters from men within the family.[99]

Women often refuse to aknowledge the rape because they are economically dependent and are convinced by male-constructed myths that rape happens 'out there', is committed by 'strangers'. They are fearful of confronting ongoing rape in their own homes by a husband to whom they are sexually and socially linked.

The mother is also accused of being responsible by abandonment. She is said to be absent or not sexually available, thus *causing* the father to look to the next available female for sexual satisfaction. So again woman is reinforced as being for man, being obliged to be constantly sexually available. Both mother and daughter relearn the message daily.

The family is structured on power relationships designed to empower the father. So 'the "collusive" mother maintains the incest secret and fails to act for the same reasons that the daughter does: both mother and daughter are powerless vis-a-vis the father'.[100] They are being trained into the definition of woman as split from other women, useful as body only, powerless within male 'protection'.

Father—daughter rape is a paradigm of the social structuring of heterosexual relationships. As Ward writes: 'Men have economic, legal, political, physical, medical and social power over women, but at the root of all these is sexual power'.[101] Data shows that there is an 'overwhelmingly heterosexual orientation in men who rape children', contrary to myths about homosexual men.[102] They exercise power over those who have little in comparison—girl children trapped in the so-called protection of the family. Through those acts, the patriarchal definition of the role of women is reinforced: powerless, submissive, and available for sexual use/abuse.

Women as friends

Woman is defined as bonding with men and children. Often trapped in the 'privacy' of the nuclear family, her time consumed in work and servicing, woman is kept separate from other women. This isolation is reinforced by the patriarchal myth that women cannot get along with each other; that because of competition for the desired male, women continually betray each other.

Men have tried to contain women's loving to service them. They need women for intimacy because of their inability to relate at an emotional level to other men. As Louise Bernikow has commented, 'football and war have been recorded cultural forms of masculine intimacy. Men care for each other in fox holes and locker rooms'.[103]

They have feared the bonding of women as a threat to patriarchal order. While isolated, women will not understand their experience of shared oppression and act collectively against it. As Mary Daly points out, 'the Female Self is the Enemy targeted by the State of War. This Self becomes ultimately threatening when she bonds in networks with other Self-accepting Female Selves'.[104]

Women have always fought to establish a sense of themselves outside the male definitions of woman. Constantly women have been told that they cannot get on together, that they are too much in competition with each other for men to be friends. As Janice Raymond writes:

Women have been friends for millennia. Women have been each other's best friends, relatives, stable companions, emotional and economic supporters, and faithful lovers. But this tradition of female friendship, like much else in women's lives, has been distorted, dismantled, destroyed—in summary, to use Mary Daly's term, *dismembered*. The dismembering of female relationships is initially the dismembering of the woman-identified Self. This lack of Self-love is grafted onto the female self under patriarchy. If the graft takes, women who do not love their Selves, cannot love others like their Selves.[105]

Raymond argues that friendship begins when a woman has an affinity with her vital Self, and that women's original and primary attraction has been for women.

In order to break these bonds, men have institutionalized the ideology of hetero-reality, thereby creating an ambivalence in women's relationships with women. Reared to compete for male attention, women learn to distrust other women. The institutionalized nuclear family creates barriers to female solidarity. These barriers are the social isolation as well as affectual isolation. Bardwick points out that women often fail to respect or trust other women because of a fused 'intimacy and distrust'.[106] Overt aggression is not sanctioned between girls and their aggression therefore becomes covert. Rejection of another girl is the easiest way to express it, and most girls learn the reality of the pattern of 'being chosen and then rejected as a special friend'. Because women are skilled in the management of interpersonal relationships, they can build and destroy them in subtle ways. Girls learn that they can be seriously hurt, and mistrust intimacy with other women.

Yet historically women have sought from their women friends the intimacy and emotional support they were unable to find from men, inside or outside marriage. Women's socialization has taught them the skills of empathy and the expression of emotion. Women show 'more intimate syles of interaction' than do men[107] and their friendships are more an exchange of confidences. Numerous studies show that women still supply husbands with emotional support, but get their own support from their women friends. For example in one study of 731 families, a third of husbands responded to their wives' stress problems 'by criticism, by rejection, by dismissal of them as unimportant or by merely passive listening'. Bernard comments that we are only now becoming aware of the costs for women of the substitution of heterosexual relationships for homosocial relationships.[108]

Woman to woman friendships are the most intimate, 'the most open, the most comfortable, the most supportive, the most reciprocal, the most "mature" of adult relationships'.[109] There are many details valued in this caring of women for women, for example, the giving of gifts which frequently occurs between women friends. Gifts 'become metaphors, describe transactions, are in themselves the language of relations ... what are these things but the laying on of hands ... we stroke, decorate, adorn and caress each other'.[110]

Bernard writes that 'there are many more female—female relationships in our society than there are female—male relationships. And the part they play not only in the lives of women, but also in the structure of the female world is incalculable'.[111] For example, women's kinship networks have always been important to them, playing an essential support role. This began to change with increased mobility, education, and work placement of women. Friendships then become primary, as pseudo-familial ties, and for women today work-site friendships are important.

The myth that women cannot trust each other serves patriarchy by reinforcing women's dependence on men. Women's relationships have always been strong, but they have been made invisible in history and there are periods when they have been strongly attacked and devalued. Women-only communities have always given men concern, often leading them to introduce prohibitions on women's friendships, as in convents (see next section for details).

In Britain in the nineteenth century, one way of seeking a community of women was through the ever-increasing nursing community. In 1873 at an annual address to the Nightingale School at St Thomas' Hospital, Florence Nightingale said 'The friendships which have begun through this School may last through life, and be a help and strength to us. For may we not regard the opportunity given for acquiring friends as one of the uses of this place'.[112]

The strength of women's communities, the increasing number of women living with women or on their own, the increasing number of spinsters and 'loose' women, the increasing economic independence of women, all caused increasing anxiety in nineteenth-century England. These freedoms for women were gained within the social context of a growing women's movement and a push for the vote. Anti-woman forces increasingly attacked women's friendships. But it was the development of sexology and the classic work of Havelock Ellis which named women's friendships for the first time as deviant and abnormal in scientific terms. In his book *Sexual Inversion* (1897) he described the female homosexual in classic stereotypic terms: using direct speech and evincing a masculine straightforwardness, freedom from shyness and a 'dislike and sometimes incapacity for needlework and other domestic occupations'. These were the forms of behaviour classified as unhealthy. Lesbianism was named as deviant and the labelling of women who have intimate relationships with each other as lesbians has since that time been used as a way of breaking female bonding.[113]

In American society, particularly in the middle and upper classes, there was also a flowering of female friendship in this period. Intimate networks were established which linked women, particularly during the transition periods of their lives. Bernard discusses this women's world as inaccessible to men; shaped by external forces (their exclusion from political, social, economic and religious power); but cultivated from the inside. 'It was a loving, supportive world' and carried a sense of solidarity, love and sisterhood.[114]

Pioneer women in America needed to sustain their women's networks over distance and time and painful isolation. Australian women settlers experienced similar feelings. Often cut off from kin ties, they established strong networks in the new land, where they were often taken to totally isolated places to live with no European women for hundreds of miles.

Carol Smith-Rosenberg has studied relationships between American women in the nineteenth century and uses detailed analyses of manuscripts and letters to trace these 'sororial' relationships. The women express passionate feelings of love and caring for each other and physical separation did not weaken these affectual ties. Letters between the women contained terms of endearment commonly found today in heterosexual love correspondence. The letters are strongly sensual. Yet Smith-Rosenberg points out that sexuality was not an issue, and both husbands and wives found the homosocial world of women to be acceptable, normal and complementary to the heterosexual world. She talks of the 'essential integrity and dignity that grew out of women's shared experiences and mutual affection'.[115] Women's friendships were what made marriage bearable for many women. They often expressed a moral superiority over men and many experienced a common revulsion against male carnality. Smith-Rosenberg emphasizes that this was not a new way of bonding for women.

In hundreds of cultures around the world and across time, women have lived in highly sex-segregated communities; spending their time with other women; developing female rituals and networks; forming primary emotional, perhaps physical and sexual, ties with other women. Such women develop visions of the world, values, and indeed, I would argue, even symbolic and cosmological systems different in highly significant ways from those of the men with whom they shared sex, food and children.[116]

Central to this world in nineteenth-century America was the core of kin relationships among women, which spread out to a network of friendships. Hostility towards other women was discouraged, and 'a milieu in which women could develop a sense of inner security and self-esteem was created'.[117] Relationships between mothers and daughters were usually harmonious, mothers training daughters for their future roles and helping them to create female networks interconnected to their own. This was a form of apprenticeship. Scorn was reserved for the male peer group. Men were indispensable in the marriage stakes, yet were alien creatures involved in their own homosocial world. Marriage was seen as a period of great adjustment and women prepared for it by close and intimate discussions within their female networks. Their friendship with women remained their most intimate social and emotional support system. The love involved in these friendships was expressed by Mary Grew who wrote in 1892 on the death of her close friend Margaret Burleigh: 'To me it seems to have been a closer union than that of most marriages. We know there have been other such between two men and also between two women. And why should there not be. Love is spiritual, only passion is sexual'.[118]

Smith-Rosenberg sees the demise of the 'rich and vital world of woman-identified women' in the nineteenth century as related to the fragmentation of feminism in the 1920s and 1930s when 'the community of self-defined autonomous women had become the subject of derision and ridicule, denigrated alternatively as lesbian, sexually repressive, and old fashioned'.[119] In some instances certainly these friendships would have had a sexual component. But passion between friends does not necessitate a sexual relationship. As Janice Raymond writes: 'One can choose to be passionate friends instead of exercising passion in a sexual/genital way'.[120]

Janice Raymond has argued that there is an original and primary attraction of women for women, which hetero-reality constantly tries to break and redefine. But Gyn/affective women are 'for women': loving, influencing and being influenced by other women. It is a term to describe women moving toward women. Raymond invigorates the tradition of women as friends, companions and lovers historically. Men, she argues, are ultimately homo-relational, using hetero-sexuality to gain support and sustenance from women, but turning their society towards men.

Raymond remarks on the social and political power of Gyn/affection, naming it as an immense force for 'disintegrating the structures of hetero-reality'. Tracing the history of women's communities such as those in convents and the Beguines, Raymond develops a philosophy of female friendship which extends our understanding of women's resistance within patriarchy. 'Behind many apparently colonised women have been women of strength and fortitude who encouraged their Selves and each other'.[121]

The passionate friendships of women can also be traced into the twentieth century. Jessie Taft, for example, was an American social scientist who wrote her dissertation on the women's movement and social consciousness in 1916. She saw the only solution to the social divisions between the sexes and the hostility of men to lie in the companionship of other women. She wrote of:

building up with them a real home, finding in them the sympathy and understanding, the bond of similar standards and values, as well as the same aesthetic and intellectual interests, that are often difficult of realisation in a husband . . . [122]

Jessie established a home with her friend Virginia Robinson and they adopted children.

Strong friendships can also develop around a passionate working relationship. A passionate commitment to a joint project, often over a lifetime, can form as strong a bond as sex and intimacy, often stronger.

Little attention has been paid to the passionate love which can exist between sisters, but for many women this bond produces a security in lifelong support and companionship which is not available through relationships. Bernikow comments that if the stories of male literature are to be believed, women would be better off without sisters. 'In this masculine vision, all women would be better off without other women, for the woman alone—motherless, sisterless, friendless—can fix her eyes solely on father, brother and therefore peace will reign in the universe. Or masculine power will have its way'.[123]

She looks at the relationship of sisters Vanessa Bell and Virginia Woolf, seeing the basis of that strong relationship as an acceptance of the need in each other to work in an area of her own without competition from the other; a commitment to each other; and a fear (with some justification) of the disruptive influence of marriage. A modern sisterly relationship which is partly public is that of feminist writers Dale and Lynne Spender.[124] Men see these relationships as dangerous to their self-interests. As Toni McNaron writes in her analysis: 'Like mothers and daughters, sisters are by definition a woman-to-woman dyad and, as such, may well constitute a threat within patriarchy'.[125]

The definition of women as unable to love each other is false, and only serves men in their demands that woman be defined as for man. The history of woman-loving has been conveniently made invisible in order to convince women of this self-identity. But even more of a problem to men are women who reject them sexually as well as emotionally.

Women as lovers

Lesbianism is a threatening reality to men, representing a direct or indirect block to their access to women. One of the reasons often cited for nineteenth-century clitoridectomies was the possibility that masturbation would lead to homosexuality between women. Men fear that women will choose lesbianism over heterosexuality if given the choice.

In England in 1921 female homosexuality was almost made illegal. It was not Queen Victoria's supposed refusal to believe that lesbians existed which stopped the legislation. Male parliamentarians were worried. 'Lesbianism was an alarming phenomenon to these M.Ps because they thought it would spread like wildfire if women even heard of it. Silence, and every attempt to make lesbianism invisible, was the only effective weapon'. A Lieutenant Moore-Brabazon suggested three ways of dealing with these 'perverts': the death penalty, locking them up as lunatics, or ignoring them. He pushed for the third method because he said it was best not to 'advertise them'. If lesbianism were not advertised, he said, 'these cases would be self-exterminating'. He continued:

They are examples of ultra-civilization, but they have the merit of exterminating themselves, and consequently they do not spread or do very much harm to society at large ... To adopt a Clause of this kind would harm by introducing into the minds of perfectly innocent people the most revolting thoughts.[126]

The Church, as an institution within which women can legitimately live separatist lives, has been constantly concerned that women should not enter sexual or even friendship relationships with other women. Religious councils of Paris (1212) and Rouen (1214) prohibited nuns from sleeping together and required a lamp to burn all night in dormitories. 'From the thirteenth century on, monastic rules usually called for nuns to stay out of each other's cells, to leave their doors unlocked so that the Abbess might check on them, and to avoid special ties of friendship within the convent'.[127]

This attitude has been carried through into the twentieth century and in *Lesbian Nuns: Breaking Silence* (1985), Nancy Manahan talks about the importance of the title 'Breaking Silence' because of the imposition on nuns of silence as a rule of convent living. Many of the nuns in that book speak of the pressure upon women not to form special friendships.[128]

In her book *Immodest Acts* (1986), Judith Brown creates a narrative around documents which tell the story of Sister Benedetta Carlini, Abbess of the Theatine Convent of the Mother of God in Renaissance Italy. Entering the convent at the age of nine, Benedetta was known by the age of twenty-three to have both religious and erotic visions. Eventually the church became suspicious of the claims of supernatural contact with God. The result was a series of ecclesiastical inquests which revealed that Benedetta and a young nun Bartolomea had been engaged in a sexual relationship for a number of years.

Brown points out that although the word 'lesbian' has appeared once in a sixteenth-century work, it was not a term commonly used until the nineteenth century. Various terms were used to describe what women actually did, including mutual masturbation, fornication, sodomy, mutual vice and so on. At one point it was called 'the sin which cannot be named', indicating a constant preoccupation by men through the centuries with what it is that lesbians actually *do*.

But one thing becomes clear through this narrative: that men could not understand sexuality between women and so assumed in their phallocentric way that women were either settling for second best because they were not allowed access to men, or were preparing themselves for relationships with men. Brantôme, late sixteenth-century commentator on the sexual antics of French courtiers, for example, wrote that 'what they get from other women is nothing but enticements to go and satisfy themselves with men'.[129] The attraction of woman for woman was not to be taken seriously, unless it challenged the authority of men.

Another reason given for ignoring lesbian activity through the centuries was the belief that women, naturally inferior to men, were merely trying to emulate them. Brown describes the various moves by various societies as early as 423 and through to the sixteenth century, to either condemn or ignore lesbian love, in the hope that women would not be enticed to experiment.

Even in the nineteenth century when Havelock Ellis and other sexologists were condemning and stereotyping lesbianism, the preoccupation with the physicality of lesbian relationships is obvious. Ellis asserted that the most common form of sexual practice was 'kissing and embracing', and that sexual contact was rare. He wrote:

Homosexual passion in women finds more or less complete expression in kissing, sleeping together, and close embraces, as in what is sometimes called 'lying spoons' . . . Mutual contact and friction of the sexual parts seems to be comparatively rare . . . While the use of the clitoris is rare in homosexuality, the use of an artificial penis is by no means uncommon and very widespread.[130]

It is characteristic of the pronouncements of sexologists and of Freud, that the clitoris is made insignificant in descriptions of women's sexuality for women. But the use of dildos, which has been a common motif of men's sexual fantasies about lesbians and features in male pornography concerning lesbians, takes pride of place. Women cannot possibly have a sexual relationship without a pseudo-phallus. Even here, men define sexuality as penetration.

Lillian Faderman has unearthed records of friendships between women in the eighteenth and nineteenth centuries, but also in the seventeenth century and in the Renaissance. She has called these relationships 'romantic friendships', seeing them as 'love in every sense except perhaps the genital', mainly because she claims that women had accepted the male definition of female as being without sexual passion.[131] She stresses the emotional life of lesbian women, defining the relationship as one in which 'two women's strongest emotions and affections are directed toward each other'. (Ruehl has criticized Faderman because her work generally excludes the sexual relationships between women.)[132] Wolff too has stressed the emotional elements involved. She says that 'all women are by nature homosexual' and stresses the high intensity of the emotional lives of lesbian women.[133]

Dealing with the Chinese Marriage Registers, communities which operated from the early nineteenth to the early twentieth centuries before being forced into exile, Raymond notes that the women were attacked as lesbians because they were 'loose women'. But many of them *were* lesbian as 'female independence does breed lesbianism'.[134] She compares these women with their Western counterparts in convents. Both had their own social, economic and cultural systems which institutionalized female friendship. In both groups the women emphasized heterosexual chastity. This was an empowering decision enabling the woman to resist and withdraw from the degradations of marriage or prostitution.

Men have named woman-loving as 'unnatural', women's sexual self being male-defined as 'naturally' heterosexual. In her beautiful poem 'Transcendental Etude', Adrienne Rich attacks the 'unnatural' label, turning it on itself, claiming separation from women to be unnatural. She stresses the original love of women for women in the mother relationship and its connection with self-love.

Only: that it is unnatural,
the homesickness for a woman, for ourselves,
for that acute joy at the shadow of her head and arms
cast on a wall, her heavy or slender
thighs on which we lay, flesh against flesh,
eyes steady on the face of love; smell of her milk, her sweat,
terror of her disappearance, all fused in this hunger
for the element they have called most dangerous, to be
lifted breathtaken on her breast, to rock within her
—even if beaten back, stranded again, to apprehend
in a sudden brine-clear thought
trembling like the tiny, orbed, endangered
egg-sac of a new world:
*This is what she was to me, and this
is how I can love myself—
as only a woman can love me.*[135]

Rich conceptualized what she called the 'lesbian continuum' to explain a range 'through each woman's life and throughout history— of woman-identified experience; not simply the fact that a woman has had a consciously desired genital experience with another woman', but meaning 'the sharing of an inner life, the binding against tyranny, the giving and receiving of practical and political support'.[136] This makes it possible for all women who are woman-identified (i.e., who put women first, at the centre, or who have intimate relationships with other women) to be included in the lesbian continuum. But it undermines the sexual element of lesbianism. As Janice Raymond points out: 'woman-identified women who are not Lesbians, while showing courage in the midst of a woman-hating society and taking other risks, have not taken the specific risk of choosing and acknowl-edging lesbian Being'.[137] While not wanting to reduce lesbian selfhood to sexual/genital contact between women, Raymond stresses that for most lesbians it does include sexual relationships and with them the risks that come from living within a homophobic world.

Haunani-Kay Trask has based her vision of feminist Eros on a feminist conception of power and love deriving from a life-protecting bond with the mother and with other women. She too considers the 'return to the mother' to be implicit in lesbian loving, and to represent a return to love. She points out the similarities in the bodies between the two lovers in such a relationship and the belief that physical gratification is more likely between women because they are more familiar with bodies like their own.[138]

Other writers have argued that the strength of lesbian feminism is in the greater relational capacities of women which increase the po-tential for growth and freedom, and the absence of patriarchal sexual politics in such relationships. Certainly both these elements are seen in the twenty-one love poems of Adrienne Rich, where the return to the mother is also a sense of 'coming home' in the fitting together of women's emotional and physical selves.

Little has been written about the sexual desire of women for women, and the stress on their emotional life does encourage a picture of lesbianism which excludes sexuality. In writing of sexual love, Bernikow comments that female love 'exists in a phallus-free world ... The language of love between women, is on the whole, not the language of—I took her ... '.[139] Shere Hite, in her survey of sexual behaviour, found that women expressed more tenderness and caring for women lovers.

Discussion of women's physical loving is best traced through novels and poetry, as in the sensuous patience of love-making between Leslie and Valerie in Marge Piercy's *The high cost of living*, or in the poetry of Adrienne Rich. Some writers argue that woman-to-woman love is a choice away from the sexual dominance of the male. 'The predatory power nexus of masculine/feminine interaction is absent, many lesbians argue, because the male is absent. And with him goes the penis-as-weapon; the aggressive, rapist encounter; the denial of female eroticism and gratification; and the pain of relational incongruity'.[140] Though some women choose lesbianism because of their political commitment, Trask, on the other hand, argues that lesbianism must grow out of sexual and emotional need and not out of politics: 'Put another way, love cannot spring from theory, no matter how persuasive'.[141]

Men fear and hate lesbianism because they fear not being 'fed' by the female. As Wolff writes: 'He needs ego support throughout his life. A lesbian who 'feeds' (loves) another woman puts him and his world into chaos; she is a rival because she takes away maternal support which should be HIS not HERS'.[142] Women, on the other hand, may dislike lesbian women because they are afraid of their own potential for a lesbian relationship which may surface and could threaten their sense of security in a hetero-relational world.

This is not to say that lesbian relationships are somehow perfect and above the normal stresses and tensions which come from trying to establish intimate relationships. In addition, there is the possibility of the merging of woman to woman. Susan Kreiger, in *The Mirror Dance* assesses the dynamics in a lesbian community in the mid-west of America. She outlines the difficulties women have in attempting to find the balance between being autonomous and their own persons, and merging with other women who are so close to oneself and so like oneself.[143] The intensity of lesbian relationships can sometimes be overwhelming. However, as Nancy Friday wrote:

It's the most natural thing in the world, for the same reason men do, women should turn to women for tenderness . . . The female breast, symbol of tenderness, is there for men to cry on, to suck on, lie on for life. But what about women?[144]

Woman's self-definition has been constructed within heterosexuality in a hetero-relational world. Men want woman's intimacy skills, emotional nurture and sexual energy directed towards them, and structure the material and ideological conditions necessary to ensure this. Women resist it through woman–woman bonding in friendship, love and sexual relationships.

Women need passion, love and intimacy. But these are not experiences isolated from the political and social condition of woman. As Janice Raymond writes, 'sex is a whole human life rooted in passion, in flesh. This whole human life is at stake always'.[145] If women love men, will men use this to justify their benefits as members of an oppressive group; can men in fact release their own desire for power to meet women on a plane of intimacy which is self-actualizing and not self-denying for women?

Conclusion: Power and Resistance

Power as it is constructed and maintained by men, underlies the definition of self imposed on women and discussed in this book. Through their control of the structures of society (economic, political, legal, religious), through language, through knowledge-construction, through ideology, and through force, men have confined woman's self, defining her negatively.

From women, men need children, unpaid labour and emotional and physical servicing. To ensure these, woman has been defined as 'nature', which is irrational and in need of control, and as suitable for mothering and servicing. She has also been defined as passive and powerless. But a corollary of power is resistance. Women have fought definitions of the self which debase and disempower them. Jean Lipman-Blumen argues that gender roles actually provide a blueprint for all other power relationships. She writes:

Gender roles are the model for power relationships between generations, socio-economic classes, religious, racial and ethnic groups, as well as between imperial powers and their colonies, and between less developed and post-industrialised societies. Power is the infrastructure of the sex−gender system. As a result gender roles are carefully guarded, or otherwise changes in this primal power relationship might spread to all other established power relationships.[1]

This chapter will analyse this infrastructure: the nature of power and powerlessness; the processes by which male power is maintained; and the ways women do resist and have resisted.

Power

Men have defined power as dominance and control, as the ability of one person to influence or control another person or groups of people. It is 'the capacity of some persons to produce intended and foreseen effects on others',[2] or 'the capacity to alter the actions of others'.[3] The term 'power' usually carries implications of force and domination; of making people do that which they would prefer not to do.

Power can be defined as constraint, as in the situation in which one group is prevented from raising issues that are not in the best interests of the power holders. So men prevent women from placing their own interests—child care for example—on the agenda. Masculine definitions of power stress power 'over' rather than power 'to'. Nancy Hartsock argues that we need to change the associative links between manliness, virility, power, and domination.[4] Considering the definitions in the Webster's dictionary of 1933, Berenice Carroll notes that words such as 'strength, vigour, energy, force, ability' are words which are synonomous with power. But in the most recent definitions of power in the Webster's dictionary, words such as control, command, and domination appear.[5]

Elizabeth Janeway stresses power as a 'process of interaction among human beings'.[6] Robin Morgan also defines power as 'a *means*: of defining, categorising, compartmentalising—relationships, people, "things"; a means of naming, even more than of accomplishing'[7]. 'Process' enters into power in the constant negotiation and renegotiation which occurs between the more and the less powerful. When power is firmly established it is unnecessary to use punishment or even reward in order to control people. The knowledge that the powerful group holds such controls is usually sufficient to evoke compliance.

Barbara Watson argues that although the definition of power as dominance covers a range of uses, 'power as ability, competence, and the closely related definition of power as energy, cover another, much wider and more interesting cluster of meanings'. Her hope is that because of women's experience of being 'other' in relation to power they 'might be able to exercise power without necessarily forgetting the lessons of the outsider':[8] that power should not represent the desire to dominate and limit the potentialities of other people.

Carroll stresses that power means autonomy as an individual and the power to assert one's will over one's body, functions and physical environment. Thus power comes to be seen as competence. She considers that slogans such as 'sisterhood is powerful' should indicate woman's claim for strength and self-determination, control over one's own life and a sense of competence and independence. She argues that this idea of power 'is bound to have a deep appeal and to have potentialities for mobilising the very energies and abilities to which it refers'.[9]

Radical lesbian feminist Charlotte Bunch has drawn distinctions between leadership and dominance. Male power is manipulative, aiming to exploit other people. A leader, however, engages with the needs and wants of followers. She uses energy and strength to encourage it in others and to create effective interaction 'rather than manipulation, domination and control'.[10]

This is *em*powerment: the ability to *empower* one's self and others rather than controlling them. Power then becomes a vital force: power over the self; over the body, the mind, work, bringing a sense of self-respect and value in the world. Helene Moglen has written that 'feminists are understandably ambivalent about accepting and using power within mainstream heirarchical structures that support relationships of domination and inequity', but 'instead of dissociating ourselves from power, we should determine ways in which power can itself be purged of its own crippling effects'.[11]

Hartsock, however, has some anxieties about these more positive and energy-centred definitions. She argues that there are three drawbacks to this approach. The first is the tendency to refuse to confront the problem that in acting upon or changing the world one can harm others. Second, these arguments stressing power as energy tend to direct attention away from the relations of domination which are inherent in current power relations. Third, often the genderedness of power 'and its importance in structuring social relations' is neglected.[12]

Tracing the power of men in four male-defined categories of society—non-technical, third-world, capitalist and socialist—Leghorn and Parker delineate three differing types of power base operating for women. These are situations of *minimal* power (for example, in Ethiopia, Peru, Algeria, Japan); *token* power (USA, Cuba, USSR, Sweden) and *negotiating* power (China). Women never really attain *equal* power.[13] The situation of token power familiar in Australia, Britain and in North America, is characterized by the ease with which advances in the relative position of women can be reversed, as seen for example in the constant threat to access to legal abortion.

The examples in negotiating power which come closest to any real control for women are notable, because women have retained or gained an element of power through having something men wanted or needed. So, for example, in traditional Iroquois society women were the controllers of the longhouses; selected the male council of elders (though did not serve on it); influenced decisions on war and controlled agricultural work. Descent was through the female line. The basis of this power was that women controlled the preparation and storage of food and property, which included large orchard areas. To travel or make war, men needed to take food with them. Leghorn and Parker point out, however, that 'this control was exerted as a form of service to men' and systematically declined on invasion with the segregation of the male and female spheres of life.[14]

Men are the powerful group, the dominant group. They have structured the world in order to maintain their power over women, for example, by their structuring of pay inequality and the sex-segregated work world. And women have been defined as the less powerful or powerless group. But women are not totally weak and powerless, rather they are *disempowered*—power and autonomy have been *taken* away from them. There is an interdependence between the sexes: but inequality should not automatically follow from this interdependence. As Lipman-Blumen writes:

When both parties to an interdependent and complementary relationship have equal control over valued resources, there is no impediment to their equal power. The crux of the problem is that every society values men's contributions and resources, their activities, privileges, even their responsibilities, more highly than women's. This greater valuation of males stems from and points to the serious power difference between women and men.[15]

But how is this power imbalance maintained?

The maintenance of male power

The omnipresence of male power is encapsulated in the words of Louisa Lawson in the *Bulletin* in 1896. She wrote:

Women are what men make them. Why, a woman can't bear a child without it being received into the hands of a male doctor; it is baptised by a fat old male person; a girl goes through life obeying laws made by men; and if she breaks them, a male magistrate sends her to gaol where a male warder handles her and looks in her cell at night to see she's alright. If she gets so far as to be hanged, a male hangman puts the rope round her neck; she is buried by a male grave digger; and she goes to a Heaven ruled over by a male God or a Hell managed by a male devil. Isn't it a wonder men didn't make the devil a woman?[16]

Male power is maintained and defined through a variety of methods: through institutions within society, through ideology, through coercion or force, through the control of resources and rewards, through the politics of intimacy and through personal power.

Institutions of male power

Social institutions are structured by men in order to reinforce the power relationship between women and men. The most obvious examples of these institutions are the family, the political system, the church, the occupationally segregated workforce and the law itself. Family structure replicates that of the state with a system of authority for men ultimately backed by force. The family is the chief institution of patriarchy, maintaining socialization of children and of women in line with the role deemed necessary for them by patriarchy. Although violence is often part of the family institution, Kate Millett has indicated that force is not always necessary because of the strength of socialization within the family: 'So perfect is its system of socialisation, so complete the general assent to its values, so long and universally has it prevailed in human society, that is scarcely seems to require violent implementation'.[17] It is within the family that the training of children for traditional sex-roles takes place. It is within the family that the economic disadvantage of women is institutionalized, within the male breadwinner and woman homemaker roles which are designated 'natural'. Within the family, children observe sexual behaviour but also the power dance between adult males and females.

School reinforces these patterns and in the playground can be seen 'the tender misogynous roots of the all-male adult world that have little need for women beyond reproduction of the species and domestic service'. Within the family and the school system, children learn that the world is different for girls and boys. They learn that the things boys do and are, are more highly valued by society and therefore 'boys pay attention to boys and so do girls . . . girls are the audience'.[18] Overt socialization is the process by which, for example, boys learn not to cry and be sissies and girls learn to stop being tomboys at puberty. They learn the appropriate behavioural characteristics for girls and boys. But they also learn through 'latent socialisation' which Lipman-Blumen describes as 'the process whereby individuals are socialised indirectly to those roles the society does not expect them to perform on a regular basis'.[19] They learn the roles of the opposite sex and *not* to perform them; although in times of crisis roles may be reversed, as in wartime. Socialization does not stop with childhood, however, and adults continue to be socialized within the family in the various phases of their lives.

Within the family women's time is consumed by the domestic role, even if they work in the paid labour force. This limits their ability and time to involve themselves in public activities through which they may gain access to institutional resources and therefore power. Through the family the sexual role of woman as well as her domestic and emotional servicing roles are defined.

Discrimination and segregation are institutionalized within the workforce, and again reinforce the power imbalance between women and men. Women are clustered in the lower-paying, lower-status jobs, and are manipulated in and out of paid work when it suits men. This means they have limited and non-continuous access to resources, such as cash, within the economy. This leads to women's lack of access to what is often called organizational or positional power.[20]

Religious institutions are another example of structures created to support male power. Women lack power and status within churches and are not in the ranks of church leaders. The church preaches wifely obedience, woman as both madonna and whore, and the 'fact' of woman's original sin, leading her to multiply in sorrow. Repeatedly the church's philosophy and ideology, represented by various interpretations of the Bible, have been used to justify female subordination.

In Australia, the Movement for the Ordination of Women has created division and dissension within the institution of the church. While challenging the structures of the church and woman's place within it, feminist Christians seem unaware of the fact that they are challenging the basic ideology and foundations on which the church depends. Men within the church fight the ordination of women with desperation because they recognize, quite justly, that the very nature of this patriarchal institution is under attack from those from whom it expects and needs conformity.[21]

The subordination of women by the church created an ideological justification for men's treatment of women within the wider society. Time and again law-makers refer to the Judeo-Christian heritage and the 'values' inherent in it. These values are used at times to refuse abortion and contraception to women because they are against the teachings of the church. They allow the beating of women because they must obey their husbands. Patriarchy needs the continuing church teaching of the subordination of women because it is a cornerstone of its ideology. What greater law-giver can society seek but a god? Woman's traditional role is firmly founded on male interpretations of biblical teachings.

A patriarchal institution which determines women's access to resources, and therefore a right to define their own lives, is the political system. Women were not denied the vote for so long for no purpose. Men have institutionalized obstacles to women's entry into the public sphere of power, by denying them the vote, by denying them the right to stand for public office or, in modern society, by making it difficult and almost impossible for women to gain political positions because they have traditional family responsibilities.

Men feel endangered when women have access to decisions made about the distribution of resources in society. Innovations which would increase women's autonomy, for example publicly funded child care, are not in the best interests of men and would threaten resource allocation towards men's needs. Stacey and Price have pointed out that the feminist movement made a radical demand in making a bid for the vote, because women demanded to be treated as persons and as individuals in the same way that adult men were treated, *unrelated* to family power or class position.[22]

It is universally true that there are fewer women than men holding political power. The percentages for countries like Australia and Britain remain small. Stacey and Price point out however that when the obstacles which are placed in the way of women in gaining access to political power are considered, it is surprising that so *many* women are succeeding.

Men prefer that women align themselves to one of the political parties, all of which are male-dominated, rather than seek women's candidates on women's issues. When women raise this possibility, they are seen as being divisive within either the conservative or the labour movements. Though myth has it that women vote conservatively, it is mostly older women who tend to vote in a more conservative fashion, while younger women are more likely to support socialist or labour parties than men of the same age.[23]

Once within the political system, women experience isolation and difficulties in combining their 'appropriate' roles as women with those of politician. As in scientific research, informal networks play important roles within politics. Many recommendations and evaluations are made informally and women need to be included in those networks and decision-making processes in order to have an impact. Cynthia Fuchs Epstein, discussing women and élites, summarizes the difficulties within the political framework:

American women find it difficult to "hang out" in political club houses where friendships are established and there is talk of whom the party will support in the next election . . . women often have to form alternative women's informal settings . . . when more women are members of an elite, they can share information among themselves, but they are still not privy to the information shared by more strategically placed men. Some of this is inevitable, so long as men and women are provided with different settings in which to relax; but whether segregated settings are kept minimal or remain defacto settings for informal decision making in the elites has important consequences for women's participation.[24]

The political system is so important in women's lives because it defines 'the ways in which power and powerlessness are distributed (and legitimated) in society and all that flows through that distribution, as well as the institutions and processes traditionally thought of as forming the political system'.[25] Control of social, economic, educational and political resources, as well as the power of the military, are in the hands of the state.

Institutions within the patriarchal social structure work to maintain a power imbalance between women and men. These social institutions are important to patriarchy for as Lipman-Blumen writes, they 'continually validate the judgements and values used in their creation and preservation. The validation is more than symbolic; it entitles the powerful to deploy institutional resources to reward those who comply and help, as well as to punish those who resist'.[26]

Ideology

As Lipman-Blumen writes:

Social institutions and the resource-distribution networks they generate are kept in place by ideology. Ideologies, or belief systems, assure us that the explanations of life that the institutions symbolise—including who is entitled to power—are correct.[27]

Patriarchal ideology validates unequal power relationships and the accumulation of resources into the hands of one powerful group, men. It also gives to that group the power to define reality for the less powerful.

Ideology is transmitted through socialization processes. But it is also transmitted through the workings of the unconscious and the manipulation of people's anxieties and fears. Lipman-Blumen outlines a number of 'control myths' about women which are part of the patriarchal ideology assisting men to control women's behaviour. She includes among these the control myth of woman as selfless which has 'deflected women from working to promote their cause, regardless of its justness'. So many times women are asked to join men's revolutionary movements and promised that their issues will be dealt with when 'the' revolution, the 'greater cause', is over. And so 'repeatedly women's efforts in major causes on behalf of others opened the window to a view of their own inequality and powerlessness. And repeatedly, promises that the gender power relationship would be addressed remain unkept'.[28]

When men do join in efforts to promote women's equality it is usually when they themselves will benefit. Examples of this include the encouragement of women in the Prohibition Movement in America, the inclusion of women in trade unions when membership boosts are needed and women's suffrage in Norway being promoted in order to advance nationalism during a time of foreign domination.[29]

Other control myths are: that men are more knowledgeable and capable than women; that men have women's best interests at heart; that the resources which men have are too difficult for women to attain; and that there is a 'naturalness' in women's current role and powerless position. Women are taught that their strengths are supposedly weaknesses in comparison to men's power. It has been one of the aims of the women's liberation movement to eliminate these control myths from the consciousness of women. Hence consciousness-raising has been an important element in helping women to understand their power and potential for power within their lives.

Women constantly fight to retain their own culture in which they define women positively and as empowered against these control myths that give women feelings of anxiety, of not fitting in, of being unlovable or not worth while. A great resource which men cultivate in men but attempt to destroy in women, is self-confidence. Without that confidence, individuals cannot attempt to gain power within their own lives, or even believe they deserve to experience it.

Male force

Women also learn to fear men. Behind their ideologies and patriarchal institutions lies the fact of male coercion and force at a state, social and individual level. Men obviously control the military and other institutions of force such as the police. Political power backed by military might is 'society's quintessential power' which men have always seen as their right.[30] Men resist strongly women's attempts to enter into these areas.

In 1981 in the US the Army called for a 'pause' in women's recruitment to reassess their role in the military with the intention of levelling it out at 10–12 per cent participation. The *US News and World Report* wrote of the concern expressed by some commanders who realized that up to 30 per cent of their female troops were in forward battle areas during division exercises. The article explained that 'what has sparked the administration's decision to re-examine the female defence role are worries that the armed forces may be becoming too dependent on women in combat situations'. Lipman-Blumen asks 'Could we tolerate this reallocation of dependency?'[31]

The reasons given for concern about women's role in combat include: pregnancy which might impair combat readiness of women; the greater size of men compared to women; the differences in male and female aggression; a loss in self-esteem which male soldiers feel when women are seen to be doing things as well as men; the suggestion that the potential enemy would perceive the army as weak with a strong women's presence. Similar arguments of course have been used in the past to show that women should not be doctors, bank managers, airline pilots and so on.

As well as state force, men use force socially and personally in the control of women through violence. This occurs within the family as well as on the street. In the private world, men are taught that they will be able to control and dominate and 'that this is the big pay off for their acceptance of an exploitative economic social order'. To be a real man they must have control over *some* men and *all* women. This violence also suits capitalism in its exploitation of male labour. As Hooks has written: 'By condoning and perpetuating male domination of women to prevent rebellion on the job, ruling male capitalists ensure that male violence will be expressed in the home and not in the workforce'.[32] Where men would be punished if they attacked employers or men of status, women are good targets because basically men will not be punished for hurting women, especially wives and lovers.

Through incest, daughters can be socialized into the female role of being sexually available for the more dominant male, while wives are socialized to accept male dominance of their children and themselves. This is extended into wife-beating, where men take it as their right to assault their wives, ensuring women understand their role as property. Rape also is the exercise of male power.

Susan Brownmiller details the use of rape in the patriarchal power structure, by which '*all* men keep *all* women in a state of fear'.[33] Australia has one of the highest rates of pack-rape in the world, often credited to the dubious 'mateship' ethic.[34] One study found in an analysis of prison records that 71 per cent of men in Pentridge for rape were there for rape by more than one man.[35] In his study of 155 convictions for counts against seventy victims Barber indicates that the rapists were all 'normal' young men.[36]

Most rapes are committed by men the victim knows and in her own home or its vicinity. Victims have been elderly women, small girls of three, and nuns—women who could not reasonably be accused of sexually arousing the man. Systematic rape is carried out in war as part of the destruction of 'property'. 'Uppity' women are treated to 'a lesson' this way. Most of the myths about rape are being dispelled, in that it is no longer seen as an accident where a woman excites a man to the point when he loses control. Many rapes involve the searching out and tracking down of the victim and include verbal, emotional and physical abuse. These acts of violence and abuse against women occur every day and they are not committed by an abstract concept— patriarchy—they are committed by actual men against living women. The rapist of a thirty-five-year-old woman in Melbourne who horribly mutilated his victim was sentenced to a non-parole period of fifteen years in 1981. His victim said: 'He said he hated women, that they're all bitches and that they should pay for it. Pay for what?'[37]

For women, living in a patriarchal society entails negotiating independence and safety, dealing on a daily level with these occurrences of woman-hating. Mary Daly comments that:

we live in a profoundly anti-female society, a misogynistic 'civilization' in which men collectively victimize women . . . It is men who rape, who sap women's energy, who deny women economic and potential power. To allow oneself to know and name these facts is to commit anti-gynocidal acts.[38]

The fact of male violence acts as a kind of terrorism on women, and controls their behaviour, teaching them how to dress, where to talk, and to keep, ironically, inside behind locked doors and windows. Few days go by without a news item to remind women of their vulnerability.[39]

Pornography is a method of social control of women through violence. Through pornography women are defined once again as object, for sexual use and abuse by men. Pornography also seeks to convince women that they like being beaten and tortured, that this is a stimulation which makes them truly feminine. Recent debates within the women's movement, particularly in America, indicate that the control myths involved in these representations of violence against women are strongly rooted within our culture and can easily be used to manipulate women when presented as 'true freedom' and 'liberation'.[40] Painfully, ideology works to reinforce male violence as a right for men and as enjoyment for women.

Hooks comments that the romances read by millions of women reinforce sex patterns that romanticize violence towards women:

Women reading romances are being encouraged to accept the idea that violence heightens and intensifies pleasure. They are also encouraged to believe that violence is a sign of masculinity and a gesture of male care, that the degree to which a man becomes violently angry corresponds to the intensity of his affection and care. Therefore, women readers learn that passive acceptance of violence is essential if they are to receive the rewards of love and care. This is often the case in women's lives. They may accept violence in intimate relationships, with a heterosexual or lesbian, because they do not wish to give up that care. They see enduring abuse as the price they pay. They know they can live without abuse; they do not think they can live without care.[41]

Male violence is also used to teach resisting women a lesson. Women are warned against upsetting the power imbalance and when they challenge it, assaults on women begin to rise. At these times the pornographic media become more sadistic in their treatment of women. Many women find it difficult to live in the reality of a world in which male violence is a major means of constraining and controlling their behaviour. They refuse to see male violence when it occurs or to accept that they too may be a victim. As Lipman-Blumen writes:

The message is clear: those who would question that men have women's best interests at heart, those who would upset the delicate, the ultimate power relationship, must be prepared to pay dearly. Rather than confront this negative picture, many still prefer to think that men simply want and love women, and that the primary source of this desire is the great beauty and sexuality that women possess.[42]

Rewards and resources

Because men are the powerful group, they control the reward structure: 'who gets rewarded for what, in what coinage, and in what amount?'[43] They can reward interpersonally, through love and support, or materially, through access to resources. The rewards that women have to offer, however, are private coinage—praise, love, reverence— currencies with little economic leverage or public power. Women's reward resources such as nurturance and emotional service are not translatable into power.

Though men can reward women by giving them access to resources, because they remain the powerful group, this access can be withdrawn at the whim of the dominant group or when women gain almost enough access to bring about real change. Resources include money, time and physical strength. For resources to be effective the person being influenced must have a need for them.

Men control economic resources; they also control the law, which is a concrete resource. For example, laws covering women's working conditions, health care, maternity and abortion can radically improve their social and economic position and therefore their power.

As power breeds and ensures more power for men, so their control of resources ensures that they continue to control resources. They are rewarded for this control by acquiring more power.

Personal power also plays a role. Assumptions behind this personal power are that men know their way about the world. They have greater self-confidence in that their place is *within* the world. A person who believes in herself and her abilities is much more likely to have success. Personal power is the acceptance 'of an individual's right to make decisions about a particular aspect of social life. Personal power is thus based on the recognition and acceptance of individual decision-making'.[44] This is legitimate power for men but not for women.

Assumptions about personal power and the right to decision-making for the individual often marks the power differential within intimate relationships. These are the *politics of intimacy*. This is the daily context of human relationships, what Henley calls the 'micro-political structure'. Here the small daily power plays operate between the sexes. These minutiae of relating, such as power in decision-making and the use of non-verbal language, form a 'continuum of social control which extends from internalised socialisation (the colonisation of the mind) at the one end to sheer physical force (guns, clubs, incarceration) at the other'.[45]

Within the basic differentiation of power according to sex, lies the experience of power relationships within other stratification systems of our society. Within varying cultures women and men may carry out different tasks and have different roles, but in all societies men have ultimate power and their resources are more highly valued than those of women. However, women experience this power imbalance differentially depending on age, class, sexuality and so on.

Upper-class women inhabit a world closer to male power and influence than other women.[46] Through their husbands they can sometimes wield influence in the larger world. But these women are usually absorbed into their husband's jobs and responsibilities, and come to symbolize the man's power position. They experience a loss of self, and are often unrecognized in their own right unless they appear accompanied by their spouse. So the power balance is 'calibrated to the type and quantity of resources available to women and men as individuals and as members of families from distinctly different social classes'.[47]

In conclusion, Hartsock brings all of these elements in the process of male power maintenance together:

Men's power to structure social relations in their own image means that women too must participate in social relations that manifest and express abstract masculinity. The most important life activities have consistently been held by the powers that be to be unworthy of those who are fully human, most centrally because of their close connections with necessity and life: motherwork (the rearing of children), housework, and until the rise of capitalism in the west, any work necessary to subsistence. In addition, these activities in contemporary capitalism are all constructed in ways that systematically degrade and destroy the minds and bodies of those who perform them. The organisation of motherhood as an institution in which a woman is alone with her children, the isolation of women from each other in domestic labour, the female pathology of loss of self in service to others—all mark the transformation of life into death, the distortion of what could have been creative and communal activity into oppressive toil, and the destruction of the possibility of community present in women's relational self-definition.[48]

A current case study in the battle to disempower women

In many areas of women's lives the battles still continue, as women struggle against the structures of male power to live with a positive sense of self, free from violent action against them. One case study which exemplifies the processes of male power imposition discussed above is the current battle around the new reproductive technologies (RTs). These are technologies which artificially create life, such as in vitro fertilization (IVF) and its variations, or which manipulate life at or around conception, for example surrogate embryo transfer, sex predetermination or genetic engineering.[49] In this field the social control of women through control over women's bodies and procreation is played out with a medical and social justification. Science and commercial enterprises (forms of institutional power) join forces with ideological forms of power (the control myth of woman as mother) in their attempt to control women's procreative power.

Men, through science and medicine, have long fought the battle for control of women's bodies (see Chapters 1 and 2). Ehrenreich and English outline clearly the history of the elimination of women healers by a rising male-dominated medical profession and the encroaching of this profession into women's lives. Birth itself now is a technological process. There is a disease model of pregnancy, where natural childbirth is seen as odd and women are encouraged to hospitalize themselves in order to give birth. There is greater technological intervention at all stages of pregnancy. Ultrasound, for example, supposedly developed in order to help women who were 'at risk', is now used routinely for all pregnancies in the United Kingdom. The Caesarean rate in Australia and America has increased up to 20 per cent in recent years with no improvement in infant mortality—the supposed reason for its introduction.[50]

Women's bodies are constantly the site of experimentation with respect to drugs. DES (diethylstilbestrol, a synthetic oestrogen) was one example of the use of women as living laboratories. It was used from the early 1940s until 1971 as a prescribed drug for pregnant women who were prone to miscarriage. Some of the women took the drug as experimental subjects and were told they were taking a vitamin tablet.[51] But there was a time-bomb effect with DES, and years later some of the daughters of these mothers are suffering cancer of the vagina and cervix at a rate higher than that of the female population of their own age. Problems have also been detected in some sons of DES mothers. There is a higher rate than normal of infertility in both sons and daughters. For DES daughters, there is an 'increased risk of spontaneous abortions, premature deliveries and ectopic gestations'.[52]

The mothers who took DES also suffer, and one study in 1984 at the Norris Cotton Cancer Centre in Hanover, New Hampshire, found 40 to 50 per cent more breast cancer in women who used DES 20 to 39 years previously.[53] Though no longer prescribed as a drug for miscarriage, DES in general is still prescribed as a morning-after pill. There is no evidence that DES does prevent miscarriage or conception, so Diana Sculley points out that women who are given it as a morning-after pill will still conceive sometimes, and rape victims who are given it as a pill to stop pregnancy may still end up being pregnant.[54] It is in the 'natural progression of science' that in vitro fertilization is now hailed as a solution to the problems of DES. DES daughters are being offered IVF pregnancies to overcome their infertility, caused by medical mismanagement and experimentation on their mothers.[55]

As the medical profession has extended its control over pregnancy, so it now desires to complete this control in the area of conception itself. As I indicated in earlier chapters when discussing the use of myths, in this area too there are telling stories. For example, the goddess Athena was 'created' when Zeus, her father, had swallowed her mother. He gave birth to Athena through his head. The 'father as creator' myth is gaining reality with the new reproductive technologies. John Yovich, an IVF scientist in Western Australia involved in the development of a new commercial company there, was described thus: '*He* produced *his* first pregnancy in 1981' (emphasis added).[56]

In her book, *The Politics of Reproduction*, Mary O'Brien discusses how the intervention of medical technology in the birthing process has broken the tradition of birth as one shared with other women.[57] As men became more involved in birthing and in the general medical control of women's bodies, the emphasis in delivery moved away from the mother toward the newborn—away from women's ritual presence and toward the relationship between father and child. Reducing the isolation of women now in maternity hospitals is represented by the presence of the father, not by a reassembling of women's ritual presence.

O'Brien goes on to discuss what she calls 'reproductive consciousness'. In her terms, the first significant historical change was the discovery of physiological paternity, which transformed male reproductive consciousness: men discovered that they delivered the seed. The second and more recent change in reproductive consciousness was triggered by technology in the form of contraception: women gained the freedom to choose or reject parenthood.

Because women labour at birth to bear children, they can be certain of their essential participation in genetic continuity; but men do not have this assurance. O'Brien includes all women in the female reproductive consciousness, whether they have children or not, since the experiences of menstruation, menopause and pregnancy all indicate the universal relationship of women to new life. Because men are not involved in the process, they experience 'alienation', which they seek to annul by their 'appropriation of the child'. They establish structures in order to do this such as the family. By law or by force, men can control women and children. Male-dominated medicine now seeks to control the meeting and development of the seed and the egg at the beginning of life.

The context of the development of the new reproductive technologies is crucial for an understanding of the threat they pose. The masculine face of science and technology brings values of control, domination and exploitation, as pointed out in Chapter 2, and science continues to resist demands that it be socially responsible for its work.

Moreover, there is an increasingly interdependent relationship between medicine and commerce which makes this social accountability even more difficult to impose. There are many financial institutions which have a power base rooted in women's biology. These include the reproductive supermarket systems being established in North America; the drug companies who are manufacturing the fertility drugs used in reproductive research; commercial enterprises being established in Australia such as IVF Australia; and at times (as in Victoria, Australia) the state acting as entrepreneur.[58] In her analysis of 'the captured womb' Anne Oakley has considered the increasing medical control over pregnancy, with the division of medicine into various specialties, representing the segmentation of women's bodies: obstetrics, gynaecology, paediatrics, neo-natal paediatrics, foetal medicine and reproductive medicine. She points out that 'womanhood and motherhood have become a battlefield for not only patriarchal but professional supremacy'.[59] But all of these sub-specialties are supported by enormous commercial enterprises, feeding them with technological assistance and drugs. So women's bodies are also a battlefield for commercial profit-making enterprises.

The alliance between commercial interests and medicine means that huge amounts of money are poured into hi-tech experimentation on women's bodies and drawn away from less glamorous work on, for example, the *prevention* of infertility. Infertility is 'big business'[60] and science literally cannot *afford* to cure infertility or to stop it occurring. To 'succeed' in business terms now, science and commerce need an ever-ready market of infertile people.

These institutions of power, science and commerce, are supported by a powerful set of ideological structures. The control myth that motherhood is essential for and desired by all women is very powerful. Women are constantly encouraged to believe this. In this context, a woman discovering herself to be infertile faces a devastating shock. She experiences feelings of grief (anger, disbelief, guilt), a jolt to the sense of self, and sometimes feelings of punishment and guilt.

The testing process to detect infertility is extremely intrusive and exhausting. It can take from six months to six years for medicine finally to diagnose that a woman is infertile. In many instances, when a woman is heterosexual, she is assumed to be the infertile partner and tests are carried out on her before the man is tested.[61] Miriam Mazor has described the testing process as 'assaultive': women are required to 'expose their bodies for tests and procedures' and to 'expose the intimate details of their sexual lives and their motivations for pregnancy'.[62]

Infertility has been used by men as justification for opening up a new area of experimentation on women. Medical science promised a technological fix for the problem, a product—the baby. But this is not a solution to infertility and is costly for women.

Reproductive technologies include surrogate embryo transfer, IVF with variations (for example donor eggs, donor sperm, or donor embryos), sex predetermination technologies and embryo experimentation.[63] I will deal only with the 'simple case' of IVF here.

In vitro fertilization is a process which can be simply described: an egg is taken from a woman, sperm from a man, the two brought together in a petri dish where the egg is fertilized and the resulting embryo is reimplanted in the woman. Though this procedure sounds simple, the practical experience of the IVF programme is painful, costly and emotionally exhausting. The process of determining the causes of infertility itself is the first stage. But with in vitro fertilization there is again a series of tests which the woman must undergo. An IVF cycle lasts for at least two weeks. Mao and Wood write:

This involves about a week of outpatient monitoring by the daily estimation of plasma 17−BETA−Oestradiol levels, the daily scoring of cervical mucus, and one or two ovarian ultrasound examinations. This is followed by another week of inpatient care, involving frequent hormonal assays, the laparoscopic collection of oocytes and, in the event of successful oocyte collection and fertilisation, the embryo transfer.[64]

In most instances, laparoscopy is carried out to 'harvest' 'ripe' eggs. It is an operation done under general anaesthetic with all the attendant risks. A fine suction needle and a guide are inserted into the uterus which has been blown up with an inert gas, and are led to the ovary where collection takes place. (Newer methods of collecting ova which do not involve a general anaesthetic—and are thus cheaper and quicker— are being developed, but these techniques involve considerable pain for the woman and it is uncertain whether they will become more widely used.) They also carry considerable risk of infection.[65]

Doctors want to implant at least three or four embryos to have a chance of success at pregnancy. So they need to collect from women more than the normal one egg per cycle which is produced. Women are thus superovulated, using potentially dangerous hormonal injections. In the early days of IVF, women were producing five or six eggs per laparoscopy. Now up to fifteen eggs per attempt have been recorded. This means that a woman's body is being asked to produce fifteen months worth of eggs in one attempt. There is some suspicion that the increase in the number of eggs produced may be related to the use of those eggs, when fertilized, for experimentation.[66]

Women are superovulated using hormone 'cocktails'. The possible damaging side-effects of the constant and long-term use of these hormones is yet to be investigated. But some dangers have already been discussed. One article in the *Medical Journal of Australia* has indicated some of the potential problems: the body's defence mechanism against superovulation is overridden, and there may be risks associated with ovarian hyperstimulation and thrombosis.[67] Other problems can include an unexpected low pregnancy rate and a higher incidence of ectopic pregnancies.[68] Henriet et al. comment that 'super ovulation is not a simple multiplication of a normal ovulation'.[69]

Superovulation may have carcinogenic effects.[70] In the *Journal of In Vitro Fertilisation and Embryo Transfer* a case report on a twenty-five-year-old woman indicated rapidly developing cancer which covered the uterus, bladder, both ovaries and appendix after such treatment. Incessant ovulation may increase the risk of cancer by not allowing the ovaries a non-ovulatory rest period or by creating rapid cell growth which might generalize. Though cautious about the links, the authors conclude that these hormones 'can act as promoters in the process of carcinogenesis'.[71] Three such cases have been reported in the medical literature.[72] Like side-effects—how many go unreported? This process used on IVF women will be generalized to other 'targeted' populations of women if embryo experimentation is approved. Commenting on the 'explosive cocktail' given to women, French doctor Ann Cabau bemoans the extension of the use of these drugs to many other groups of women, including those who have irregular menstrual cycles, successive abortions or—husbands with defective sperm.[73] Women again are being used as living laboratories.

But superovulation is not intended to stop with IVF women. Researchers keen to do embryo experimentation want access to the general population of women. Women who undergo sterilization operations or other kinds of operations in the pelvic area could be superovulated and have their eggs 'harvested' while the operation was taking place. This is already occurring in the UK. Dr Alexander Templeton and his team at the Medical Research Council's Reproductive Biology Unit in Edinburgh have grown embryos for research, having gained the eggs from women being sterilized.[74] At least three further centres in Britain are also using 'donated' eggs for experimentation.[75] To date there are no guidelines in any country to guard against the possible coercion of these potential 'donors'. There seems to be no guard against eggs being taken without the consent of patients.

Through reproductive technology, women are being dissected and dismembered. The language of medical research talks about 'uterine environments' and 'harvesting eggs' as if women are body parts and not whole persons. This is exemplified in the term 'surrogate mother' which is a misnomer. The woman is not a surrogate but is in fact the birth mother. Yet she has now been described by one writer as the 'illegitimate mother', making it easier for patriarchy to tear the woman from her child.[76]

IVF itself also involves enormous financial cost both to the individuals involved and to the community in terms of the use of facilities, expertise and funding. But the most notable thing about it is that it is a false promise and a failed technology.

The success rate for live births in Australia is between 5 and 10 per cent (though in some states it is as low as 4 per cent). That is, for every hundred women who go on to a programme about ten of them will take home a baby. Evidence from the Perinatal Statistics Unit at Sydney University indicates that the pregnancy rate is higher, but only 68 per cent of the pregnancies result in live births (many of them multiple births), so women who are already experiencing the grief of infertility now have to deal with the grief of a failed pregnancy.[77] There is a 25 per cent spontaneous abortion rate; 27 per cent of the babies are born prematurely; 22 per cent of the births are multiple pregnancies with all the attendant problems; and 36 per cent of the babies have a low birth weight rate, four times that in the normal population. The 20 per cent Caesarean delivery rate in the normal population, which women are becoming more and more concerned about, fades into significance next to the 43 per cent Caesarean rate on IVF programmes, carrying with it a two to five times higher risk of maternal death. IVF is basically an *unsuccessful* technology. But the pictures of happy babies and mothers which are shown to women through another institution of patriarchal power, the popular media, do not relate these facts.

It is only recently that the views of women on IVF programmes have become known. In recent studies, women indicate the lack of dignity involved in the process. As one woman wrote: 'It affected my self-esteem. It made me depressed for a least twelve months. It affected our relationship. It affected my sexuality. I felt powerless'.[78] Many of the women comment on the lack of information given to them about the programme and in particular about the lack of success. Just as the general public has been impressed by pictures of happy babies and happy couples, so too women on the programme have been influenced by this media presentation. The complaints of women on IVF are similar to those of women in relation to other medical procedures. They feel powerless and unable to ask questions. They complain about the lack of feedback and lack of support that they get from medical staff. They feel that information is hidden from them. As one woman said:

I would really have liked to have gone back and talked to [my gynaecologist] after it didn't work out, but as [the IVF scientist] says, 'You're history, we're on to the next one, we haven't time for you now, we want to get on with it'.[79]

It does not seem that long-term clinical trials are being conducted on the drugs given to the women. Doctors repeatedly advise that there are no side-effects to clomiphene citrate, one of the hormones given. But many women have reported unpleasant side-effects. One Geelong woman said that she felt 'depressed, spaced out, lethargic and over emotional' on it. After six months, she had chronic diarrhoea, nausea, headaches and depression. She had to have an ovarian cyst removed.[80]

A Dutch woman wrote that the drug had negative mental and physical effects. After a year on it she said: 'I couldn't see sharply anymore. I saw lights and colours and I felt kind of strange/funny inside my head. I also suffered from a pain in my belly which dragged on and on. Emotionally, I wasn't stable anymore'.[81] Women speak of the refusal of doctors to *hear* them when they speak of these symptoms. They are invisible and voiceless, just as women are who suffer from Depo Provera, IUDs and the pill.

One woman on a programme who was a doctor herself said that the hormones had induced enormous emotional turmoil even though she was told there would be no side-effects. Another woman said:

The professor tells us that according to the labels in his books, they don't have side-effects. Once someone comes out of this and is brave enough to say 'you get side-effects' other women say so too. I think that's what he's worried about—that side-effects are catching.[82]

Many of these comments are resonant of the historical relationship between women and medicine; their anxieties and their concerns are not given validity. As one woman pointed out: 'Our generation were guinea pigs for the Dalkon Shield, and now we're guinea pigs for this new form of modern technology'. Patients are unable to wield any power within the traditional doctor/patient relationship, they are often too involved to consider the wider implications of IVF and are anxious that any questioning of the process will find them cut off the programme. As one patient wrote:

We as patients, are not in a position to comment objectively about many IVF issues. Always we are conscious of the fact that we are in the 'compromising' position. For most couples our dearest wish is to have a child so we do not publicly complain about the endless experimental procedures, the dehumanised method of treatment, the pain, cost and emotional strain that is an integral part of IVF. I have known some to complain, but only to incur the wrath of the IVF team.[83]

Ironically, women wishing to gain some control back over their fertility, once they have faced the knowledge that infertility has meant a loss of control in one aspect of their lives, actually yield control on reproductive technology programmes. Seeking to gain power, they actually lose power.

Women need to be wary of alliances between the state, medicine, and commerce. But they also need to be wary of the control myths that would induce them to undergo such procedures. The guilt which would result from anything being wrong with the child is enough to drive a woman to undergo all manner of medical abuse if she is convinced that it will safeguard her child. Self-sacrifice is part of women's self-definition, providing convenient living laboratories for medical science.

But the ultimate solution in terms of controlling what woman is and her procreative power is on the horizon. This involves the possibility of transsexual men becoming mothers. Dr Roy Hertz has had success with transplanting fertilized eggs of a female baboon into the abdominal cavity of a male baboon. It seems that a fetus may attach itself to any site in the body which is rich in blood and nutrients. In May 1979 Margaret Martin gave birth to a baby girl, having undergone a hysterectomy eight months earlier. The fertilized egg had lodged in her abdomen, on her bowel, where it received enough nutrients to grow to term without the aid of a uterus. In about one thousand documented cases, a fertilized egg has worked its way into the abdominal cavity of a woman which can expand to accommodate the growing foetus. Approximately 9 per cent of these women have actually given birth to healthy babies. The mother runs an enormous risk during this process and can often die from a massive haemorrhage. But the possibility of men bearing children has already been seen in this precedent.[84]

One possibility discussed is the implantation of a fertilized egg in a male abdominal cavity, administering hormones to the 'male mother' to 'mimic that of a pregnant women' and delivering the baby through a laparotomy. It has even been suggested that a woman could conceive a fertilized egg which would be flushed out of her womb and implanted in the man. There have been suggestions that transsexuals could have their sperm frozen before their conversion operation and use a donor egg with their own sperm. They would then be both mother and father to the child—the patriarchal dream.

Socially and psychologically the softening up phase for this scenario has already begun. Articles are constantly appearing in the popular press which stress the 'rights' of transsexuals to babies. In July 1984 a group of at least six male to female transsexuals requested admittance to the IVF programme at the Queen Victoria Medical Centre in Melbourne.[85] It is clear that what transsexuals want is to fulfil the stereotyped view of 'woman'. One article in 1984 said: 'Phillip McKernan wants to give birth to prove something to himself—that he has finally made it as a woman'.[86] In a 1986 article the issue was raised again by Professor William Walters, who ran the Transsexual Conversion Clinic at Queen Victoria Hospital and was associated with the Monash IVF team. He could quite understand the demand of transsexual Estelle Croot to have a baby. Said Walters: 'It is a natural corollary that they should want to have children', calling up the control myth of the 'naturalness' of motherhood for women.[87] Estelle himself said: 'I am a woman. And like any woman I want to feel complete. I want to be fulfilled and for me that means having a baby'.[88]

Janice Raymond has argued that transsexualism represents the final colonization of women.[89] Through transsexualism, which is mainly 'male-to-constructed-female' sex change, men are able to possess women's bodies, women's creative energies, women's capacities. These are the most 'feminine' women. It is a woman made by a man to be as feminine as man deems fit. As one transsexual said: 'Genetic women are becoming quite obsolete'.[90] And soon, the male-created woman may be able to have the male-created baby; a child that is neither born of woman nor connected with woman in any way. Though these will be few in number, it certainly opens the possibility to other men.

The new reproductive technologies pose an enormous threat to women in that they are taking away from them what for some is their last power base. Through that process women are used as living laboratories, as experimental subjects with no redress against the medical abuse and dismemberment of their bodies. Within this process men, through patriarchal institutions of power such as science, the law and the economy, use ideology and control myths to forge ahead and gain control over women's bodies through control of conception itself. And so the new reproductive technologies join other strategies outlined in this book for the social control of women and woman's self-definition.

The nature of powerlessness

Knowledge, which is male-controlled, has dealt with analysis of the power*ful*. It rarely focuses on the position of the power*less* or the disempowered. Rebellions and resistance of the weak have not been documented because keeping a record would be dangerous: it might inspire others to revolt.[91] Hooks has written that women do not need to 'obtain power' before they can effectively resist sexism: 'women, even the most oppressed among us, do exercise some power'.[92] Exploration of the nature of powerlessness indicates that the relationship between powerful men and less powerful women is more complex than that implied by the terms 'powerful' and 'powerless'.

The powerful cannot govern without the powerless. Inherent in the relationship is a fear of the less powerful by the powerful; a fear of women by men. Some of these specific male concerns have been discussed throughout this book. Women have been feared partly because they have been seen us unpredictable. The abnormality of women or of any group of 'weak' people, 'make the weak incomprehensible to their rulers' and 'their conduct can't be predicted by the rules of the powerful'. They are therefore seen as bewildering and incoherent.[93]

The powerless fear the totality of power of those oppressing them, including their potential for destroying them. They understand that any rights and power which they have attained can be taken away by the dominant group. So women live with the understanding and fear that their freedom can be even further curtailed. 'Rights are temporary gifts, granted if the powerful think it desirable but withdrawn at pleasure'.[94]

However, it is the disempowered, women, who know best how power works. Even when women do not consciously express or understand this they know it: 'their own everyday lives rub the awareness into them'.[95] In order to survive, women have learned very well the ways of men. They live more intimately with the oppressor than other oppressed groups. Women understand the way power operates in the world and show this by the way they pass on survival techniques to their children. But women are generally unaware that this understanding of the processes of male power is one of their strengths and could be utilized more effectively in the sex power struggle.

But why do women comply? Basically they do so because at the back of their minds is the knowledge and fear of the potential and actual use of force by men. This may be the force of the law, or it may be physical force.

Women also comply because many are convinced by ideological control—control myths—that male domination is inevitable, even natural. Discussing the experience of less powerful groups, Lipman-Blumen points out that control myths include beliefs that the dominant group is more knowledgeable and capable; that it has the interests of the weaker group at heart; that the dominant group controls valuable and otherwise unattainable resources; that the disparity in resources is so overwhelming that efforts to change it—even by physical force—are doomed to failure; and that the weaker group, if it behaves properly, can vicariously share the dominant group's power.[96] So women often accept their weaker role as inevitable, a result of their inadequacy, and as the best possible situation under the circumstances.

But men, the powerful group, need and want things from women. They depend on women's labour, both in the workforce and in the home. Resources which women do have access to, such as their emotional management skills and their skills within the family, are crucial resources which men need access to. Men need access to women's bodies and to their reproductive functioning and they will fight in the courts to ensure that they have it.

The powerful also need confirmation of their authority and the consent of those ruled. Force will only compel obedience but not committed obedience. The powerful need women to accept and sustain their inter-relationship. Men want women to tell them that they are happy with the situation that they are in, that it is a just situation; and that they love men. Men want women to acknowledge that they have the right to rule.

So women do have some crucial bargaining areas: in their ability to withdraw reproductive services, emotional services, physical labour, domestic labour, sexual labour, and their consent to being defined as the powerless, thereby verifying man's right to power.

Women also have what Janeway calls the power of *disbelief*. Wrong, in his analysis of power, wrote:

Helpless to resist coercion, fearful of punishment, dependent on the powerful for satisfaction of basic needs and for any opportunities for autonomous choice and activity, the powerless are inescapably subject to a will to believe in the ultimate benevolence of the power holder, in his acceptance in the last analysis of some limits to what he will demand of or inflict upon them, grounded in at least a residual concern for their interests.[97]

The powerful need those ruled to believe in them and believe in the justness of their position. Disbelief signifies a lack of sanction of the authority of the ruler by the ruled, and it makes men anxious and fearful, but also sometimes violent.

Women can also exercise the power of disbelief with respect to the self of woman as defined by man. As Janeway writes:

Ordered use of the power to disbelieve, the first power of the weak, begins here, with the refusal to accept the definition of one's self that is put forward by the powerful. It is true that one may not have a coherent self definition to set against the status assigned by the established social mythology, but that is not necessary for dissent. By disbelieving, one would be led toward doubting prescribed codes of behaviour, and as one begins to act in ways that deviate from the norm in any degree, it becomes clear that in fact there is *not* just one right way to handle or understand events.[98]

But men will work to divide women from each other and to eliminate this disbelief. The power of men lies in their ability to define and therefore control perspectives on the world, as well as social structures which contain women materially. Because men have set up the control myths, women need to work very hard to sustain their disbelief in the definition of woman given to them by men.

The powerless also need a collective understanding of their common condition and trust in their own perspectives and visions. They need collective action in order to change that condition. Through collective political action and through consciousness-raising techniques, women have developed a sense of female identity and solidarity.

Women have power over the domestic. This is often defined negatively by patriarchy but it is the sphere in which people live out their intimate, daily and personal lives. Through the domestic women can often exert influence if not power. There is often an interpenetration of the public and private spheres, when one group tries to influence the events in the other group's arena. So, for example, women's contraceptive behaviour can influence the development of extra schools.[99]

In her analysis of the powers of the powerless, Carroll has a large list of varying forms of power, including innovative power, inertial power, socializing power, legitimizing power, expressive power, the power of collective and co-operative actions. The titles of these forms of power are self-defining and include the power to resist by inaction and the power to resist by evasive action.

As a subordinated group women have learned interpersonal skills to gain them influence, if not power. Women do the emotional servicing work both in personal relationships and in society in general. These skills often lead women's power to be labeled by men as manipulative, but it is really management power. Women *manage* social situations, and this does give them access to influence. Lipman-Blumen comments that when the dominant group controls a society as men control our society, they rely on 'macro-manipulation through law, social policy and military might'. Women, who are the less powerful then become adept at 'micro-manipulation, using intelligence, canniness, intuition, interpersonal skill, charm, sexuality, deception, and avoidance to offset the control of the powerful'.[100]

Women, daily, learn and use survival strategies. So, for example, women become skilled at interpreting body language and non-verbal communication. Non-verbal communication (tones of voice, body posture, eye contact or lack of it, smiling, gesture and body positioning) plays an important role in communications with other people. Nancy Henley contends that these messages are part of a power game in which men communicate their 'superiority' and right to dominate space, while women communicate their acceptance of a subordinate position.[101] Generally women and men use non-verbal cues differently: women are better senders and receivers of non-verbal information. It is one of the skills developed by the less powerful.

If men are such a powerful group, what holds them back from crushing women totally? Basically, men need women—for reproduction, labour, domestic servicing, and sexual and emotional labour. They also need approval from women and a justification of their more powerful position. Men as a social group use brute force against women regularly in our society. They also know that it is not productive on the whole to use such force against women too often. As Lipman-Blumen writes:

Moreover, if what the powerful want from the powerless is love, affection, obedience, and production—either in the form of new human life, personal goods or service, or economically marketable commodities and services—they know they cannot get these by force. The truly oppressed produce little; they are too worn down.[102]

Because women are living in a situation structured so that they are separated from other women, they also come to depend on the companionship of men. They live together in social and sexual intimacy and so look to one another for reassurance in the world and for emotional and economic support. Such structured intimate living between women and men fosters growth of social bonds which ease the power struggle between individual women and men and between women and men as social groups.

Women fear the unknown. They are anxious that if their own resources were mobilized and used to change the power balance, the fabric of society might be changed in patterns which they could not control. This is particularly true for anti-feminist women who struggle to maintain the existing power relationships because they fear negative outcomes of any change. As Andrea Dworkin writes: 'the powerless are not quick to put their faith in the powerless. The powerless need the powerful'.[103]

So the relationship between the powerful and the powerless is not simple. When the odds are considered as they stand against women, their autonomy and positive self-definition, it is a credit to women's resistance and strength that they have not become totally downtrodden or obliterated in the sense of human purpose and integrity. Women have learned to glean strength from their position and to develop strategies for survival. In a comprehensive summary statement which draws together many of the threads discussed above, Janeway writes in *Powers of the Weak*:

there is a kind of courage that is very familiar to the weak: endurance, patience, stamina, the ability to repeat everyday tasks every day, these are the forms of courage that have allowed generations of the governed to survive without losing ultimate hope. The knowledge of one's own vulnerability, the choice of restraint in the face of provocation, the ability to hear one's self described as unworthy without accepting the stigma as final—that takes courage of a high order. We do not want to lose it, simply to supplement, for it's still a source of strength when the time comes to be patient no longer, when direct confrontation with the powerful for independent aims must be risked if not sought. The weak who come, or are driven to autonomous action, which is by definition unexpected and out of character, soon discover that once is not enough. A single public act is easily ignored. A short series may be observed, but misinterpreted. An event must be repeated again and again if it's to be understood; and if it isn't, it will not have the result that's expected. There is nothing so valuable as endurance, patience, and stamina when one must do, over and over, the self-same deed before this knock on the door of public attention is heard. The courage of the weak sustains persistence for positive goals when it is used positively, just as it supports resistance when that is the only possible aim.[104]

Woman's resistance

Anne Summers has written that women in Australia are

forced to eke out a precarious psychic and physical existence within a society which has denied them cultural potency and economic independence and hence has prevented women from being able to construct their own identities or from having more than a very restricted choice about what they can do with their lives.[105]

But within patriarchy, women have always resisted and fought this oppression and deterministic thinking. The existence and persistence of that struggle has been conveniently hidden from history so that women will be unaware of it and convinced that their situation is natural, and therefore immutable. Karen Horney, a psychoanalyst, writes:

At any given time, the more powerful side will create an ideology suitable to help maintain its position and to make this position acceptable to the weaker one . . . It is the function of such an ideology to deny or conceal the existence of a struggle. Here is one of the answers to the question . . . as to why we have so little awareness of the fact that there is a struggle between the sexes. It is in the interests of men to obscure this fact.[106]

Women are not the passive victims of history, but have been active agents of social and personal change. As Gerda Lerner writes: 'the true history of women is the history of their on-going functioning in that male-defined world, *on their own terms.*'[107]

In her discussion of Australian women during the 1930s depression, Stone explores this point. Using primary source material, she documents the militancy of Australian women in a variety of occupations and groups. Women were not passive receptors of inequality during this period but militant activists; 'leaders in grass-roots struggles and fighters in their own right'. She concludes that this picture 'offers a basis for optimism, a basis for a strategy to build on women's struggles rather than weaknesses'.[108]

The invisibility of women's resistance is part of the general invisibility of women's heritage within history, which Marie Louise Janssen-Jurreit writes is, 'nothing short of a stag party'.[109] She quotes Golo Mann on the importance of history for a self-identity:

Just as the individual is constituted of all the experiences of his life from earliest childhood on, the past belongs to the present 'self' of a people; what memory is to the individual, historiography is to the people ... If civilisation were deprived of any conscious contact with its history, it would not remain in one piece for long ... in the process of comparing ourselves historically, we grasp our own origins.[110]

Deprived of a conscious and positive story of herself through the centuries, woman's sense of continuity in identity with other women has been obstructed. Without a collective identity and a shared awareness of the past, a social group 'suffers from a kind of collective amnesia, which makes it vulnerable to the impositions of dubious stereotypes, as well as limiting prejudices about what is right and proper for it to do or not to do'.[111] The continuous representation of the history of men as the history of all people has denied women their place in the making of society and the creation of thought; has devalued their role within social movements; and has placed a value on the milestones within history which men, but not necessarily women, value. So it is difficult for a woman to 'orient herself to bring her personal experience into continuity with the past'.[112]

Within their past men look for guidance, confirmation and inspiration. Women too need their past for these purposes: to understand their past and present position, and to build on the achievements and learn from the failures of the past. Dale Spender points out that when women discover how men have 'doctored the records' and reclaim their past 'Then we *do* feel that self-esteem, that confidence, that sense of liberation that men take for granted in encountering their own past and finding themselves central ...'[113] It is part of women's oppression that their history has been silenced. Until they can reconstruct it and sustain if from one generation to the next, oppression is fortified.

Women need to resist the processes of power outlined above through which man maintains dominance. They need to create new ideologies and value systems which are woman-affirming. This is part of the role of feminism and the women's liberation movement, which is a resistance movement.

One line of battle for women has been fighting from within institutional power bases. An example of this is the increased number of feminists working in the public service. This is particularly true in Australia, where they have earned themselves the title 'femocrats'.[114] Since women have started entering into the structures of power there have been debates within feminism about whether it is possible to work within a patriarchal structure and still maintain the values and goals of feminism. Many of these women attempt to use what Hester Eisenstein has called 'feminist judo', that is 'cleverly placing yourself so as to use the overwhelming weight of state power in your favour'.[115] But Sue Wills has pointed out the enormous pressures on women within the bureaucracy to conform to the view of bureaucrats about the desirable limits of change.[116]

Some feminists believe in the value of working for pro-women rules of the state, such as affirmative action legislation. Others argue that those are just moves by patriarchy to buy off the women involved and to co-opt both their energy and their ideas. Jocelynne Scutt has outlined the difficulty of trying to introduce radical rape laws in New South Wales in 1981. She traces the history of the battle to convince the government to allow women's participation in the development and drafting of the Bill. But the most disappointing thing was that when the Women's Electoral Lobby Bill finally looked like being accepted by the government, it was femocrats who weakened its potential by watering down the Bill. So, she argues, women in the bureaucracy were manipulated into being used against each other. She asks:

Was their failure to be supportive, and worse, their active subversion of feminist demands, the result of genuine ability to sum up the position correctly, so that for the sake of what reforms were possible (in their view) they tailored the demands to suit? Or were the bureaucratic games that were played the result of intoxication with the 'power' they thought they wielded in advisory and governmental positions? Was there a desire to see the reform of the rape law as theirs alone, rather than the result of a massive movement, of cooperation between thousands of women? Are women equally in danger as men are of falling into the trap of individualism, rather than continuing to maintain the view that collective action brings about feminist gains?[117]

A woman can become isolated as she enters the realms of male power if she begins to model herself on the masculine universe, seeing this as a valid way of operating. Women need to maintain feminist values of woman-centredness and accountability to avoid co-option.

For all the difficulties and pitfalls, Eisenstein argues, it is still possible to work through the state. She holds that it is inaccurate to say that 'the State is male', but it is more accurate to say that 'up to now the State has been male'. So a foot in the door becomes a first requisite for change. 'Affirmative action and other "reformist" measures on behalf of women may not be sufficient. But they are certainly necessary.'[118]

Women need to maintain the political awareness that newly gained power will only advance women's position if it is consciously used with that purpose in mind. Considering the institution of the economy as an area of power, Bell Hooks delineates very clearly the dangers of the argument that all women need is *more* women within existing structures. She argues that women cannot gain much power as it is defined by the existing social structure 'without undermining the struggle to end sexist oppression'.[119] Women who are 'successful' economically in the capitalist system show a lack of understanding of 'the process by which individuals gain money and power'.[120] They gain power by supporting and thus perpetuating a dominant ideology of the culture, which necessitates exploiting people. Hooks suggests that women should exercise power in the economic sphere through their power of consumption, where women could wield power through collective boycotts and various buying programmes.

Joan Rothschild has also explored the dilemmas of women and power. She argues that 'a feminist interdependent concept of freedom is incompatible with the competitive, dominant power model, and more in tune with a co-operative, societal model of power', thus stressing the feminist understanding of the relationship between the personal and the political. *Process* is as important as the goals aimed for: 'there can be no separation of the revolutionary process from the revolutionary goal'. Striving to be participatory, co-operative and non-hierarchical, 'women must continue to pay attention to the *way* to organise, the *way* to relate to each other, the *way* to allocate tasks and share knowledge—and thus power—the *way* to make decisions.'[121]

Charlotte Bunch sees reform as part of the movement toward revolutionary change. But reform must not acquire power and change for a privileged few. It should be evaluated on five criteria: does it materially improve the lives of women and if so, which women; does it build self-respect and strength in women at both the individual and group level; does it educate women, enabling them to challenge the system; and does it 'weaken patriarchal control of society's institutions and help women gain power over them?'.[122]

Concerted collective action is an essential part of women's resistance and it is collective actions which have often brought about institutional power changes for women. Connected with collective action is activism through networking. The collective action and networking of the International Network on Female Sexual Slavery and the International Network of Resistance to Reproductive and Genetic Engineering are examples of women educating for activism against violence to women.[123] Women's health centres and the development of refuges and rape crisis centres are other examples of collective actions of resistance.

The influence and power of the women's liberation movement, a movement of resistance to patriarchy, is felt globally. Robin Morgan's book *Sisterhood is Global* is witness to this.[124] In it she draws together contributions from feminists from seventy countries, the majority of which are Third World countries. Contributions to the book highlight the commonality of women's experiences of oppression, and the resistance of women to patriarchy in each country.

Hester Eisenstein has said that we must assess feminism by taking 'a very long view, and by measuring the distance that women have come against the distance that they have had to travel'. She argues that a commitment to optimism itself is 'a feminist political stance'.[125] It is women's resistance to patriarchy which gives feminists this optimism.

So women can resist patriarchy through entering structures of power with the intention of changing them, through refusing to be domestic, sexual and emotional servicers of men, through collective action on behalf of women and through developing women's awareness of the operations of patriarchial oppression. But change also needs to begin within the consciousness of women themselves. Hooks has pointed out that if women were to rule society, 'they would organise it differently only if they had a different value system'.[126] Women also need to reconceptualize power itself. Bunch writes:

In order to end patriarchy and create a new society, women must have power. We must have power in all spheres—political, economic, and cultural—as well as power over our own beings. Since we seek power as a means of transforming society, we must also transform power or find new ways of exercising power that do not duplicate the oppressions of today. We must discover how women can build our own strengths, create these new forms, prepare for, and gain such power.[127]

Women too must unlearn the socialization and the negative definitions of woman with which they live within patriarchy at the cost of selfhood. And here women can utilize Janeway's concept of '*the ordered use of the power to disbelieve*' (emphasis added).[128] This includes the disbelief in masculine definitions of power. It includes an emphasis on self-empowerment, a sense of mastery and the breaking of control myths which define woman's being. Women must learn to share this disbelief in the patriarchal definition of their selves and to create trust in one another and in their own judgement. If they trust their own judgement they will begin to distrust the rulers. They will begin to mistrust the rules. They will begin to mistrust and redefine woman. As Janeway writes:

Disbelief and mistrust shared with others act as a protective shield beneath which a new trust can grow, trust of one's self and one's fellows. Coming together sets up a bridge which stretches from dissent to positive action.[129]

Women need to fight patriarchy at all the sites of oppression. And we cannot allow the political battle to 'stagnate at the level of personal change'.[130] But no battle takes place, no resistance grows, unless women throw off the shackles of the male-created identity of 'woman'. Women need to reclaim their bodies, their minds, their creativity, their relationships with others, their love of women, their work and their own sense of power. There is a valuable lesson in the contemporary women's movement: that 'change in consciousness and in the social relations of the individual is one of the most important components of political change.'[131]

For women, to be self-identified, not dependent on others' definitions of their selves, to be full of self-knowledge and self-respect, should serve as a solid base for action in their own interests and in the interests of all women. When women understand how men have tried to define and control woman's identity, they understand the resistance which has already come from women's energy and the constant struggle against that false yardstick. Fanny Wright in 1829 upbraided us to begin first with women and 'to analyze what we know, how we know it and whether we should begin to know very different things'. The challenge still remains to 'remove the evil first from the minds of women, then from their condition, and then from your laws'. Women need to practise disbelief and to mistrust the false self of woman which has been male-created and male-serving. Impassioned with disbelief, they must watch that new trust grow among women, so that women alone define and create woman herself.

Notes and References

Introduction

[1] Monique Plaza, ' "Phallomorphic Power" and the Psychology of "Woman". A patriarchal chain', *Ideology and Consciousness* 4, 1978, p. 5.

[2] Gerda Lerner, *The Female Experience: An American Documentary*, Bobbs-Merrill, New York, 1977, p. xxvii.

[3] Michelle Zimbalist Rosaldo, 'Woman, Culture and Society: A Theoretical Overview', in Michelle Zimbalist Rosaldo and Louise Lamphere (eds), *Woman, Culture and Society*, Stanford University Press, Stanford, California, 1974.

[4] Ruth Bleier, *Science and Gender: A Critique of Biology and its Theories on Women*, Pergamon Press, London, 1984, p. 162.

[5] Bell Hooks, *Feminist Theory: From Margin to Center*, South End Press, Boston, 1984, pp. 5–6.

[6] Jo Freeman, *Women: A Feminist Perspective*, 2nd edn, Mayfield, Palo Alto, California, 1979.

[7] Charlotte Bunch, 'Not for Lesbians Only', in *Building Feminist Theory: Essays from Quest, a Feminist Quarterly*, Longman, New York, 1981, p. 194.

[8] Hooks, op. cit., p. 58.

[9] Charlotte Bunch and Coletta Reid, 'Revolution Begins at Home', *The Furies* 1, 4, 1972, pp. 2–3.

[10] Nancy Hartsock, 'Feminist Theory and the Development of Revolutionary Strategy', in Zillah Eisenstein (ed.), *Capitalist Patriarchy and the Case for Socialist Feminism*, Monthly Review Press, New York, 1979, p. 71.

[11] Anne Summers, *Damned Whores and God's Police: The Colonization of Women in Australia*, Penguin, Melbourne, 1975. The term 'Aboriginal' is used here to indicate the original black populations of Australia, acknowledging that within Australia today Victorian aboriginals use the term 'Koori', and other groups also have specific identifying titles.

[12] Bobbi Sykes, in Robyn Rowland (ed.), *Women Who Do and Women Who Don't, Join The Women's Movement*, Routledge & Kegan Paul, London, 1984, p. 68.

[13] Adrienne Rich, *On Lies, Secrets and Silences: Selected Prose 1966–1978*, Virago, London, 1980, p. 306.

[14] Ngahuia Te Awekotuku, in Rowland, op. cit., p. 226.

[15] Hester Eisenstein, *Contemporary Feminist Thought*, Allen & Unwin, Sydney, 1984.

[16] Michèle Barrett, *Women's Oppression Today: Problems in Marxist Feminist Thought*, Verso, London, 1980, p. 256.

[17] Zillah Eisenstein, op. cit., p. 32.

[18] Hooks, op. cit., p. 65.

[19] Lisa Leghorn and Katherine Parker, *Women's Worth: Sexual Economics and the World of Women*, Routledge & Kegan Paul, Boston, 1981, p. 254.

[20] Rich, op. cit., p. 145.

[21] Catharine MacKinnon, 'Feminism, Marxism, Method, and the State: An Agenda for Theory', *Signs: Journal of Women in Culture and Society* 7, 3, 1982, p. 520.

[22] Dale Spender, in Rowland, op. cit., p. 212.

[23] Rebecca Albury, 'Reproductive Technology and Feminism', *Australian Left Review* 89, Spring 1984, pp. 46–55.

[24] Anna Coote and Beatrix Campbell, *Sweet Freedom: The Struggle for Women's Liberation*, Picador, London, 1982. p. 240.

[25] Diana Leonard, 'Male Feminists and Divided Women', in Scarlet Friedman and Elizabeth Sarah (eds), *On the Problem of Men: Two Feminist Conferences*, Women's Press, London, 1982, p. 158.

[26] ibid., p. 161.

[27] Dale Spender, in Rowland, op. cit., p. 213.

[28] Robyn Rowland, 'Women's Studies Courses: Pragmatic and Political Issues Concerning their Establishment and Design', *Women's Studies International Forum* 5, 5, 1982, pp. 487–95.

[29] MacKinnon, op. cit., p. 543.

[30] Sheila Ruth, *Issues in Feminism: A First Course in Women's Studies*, Houghton Mifflin, Boston, 1980 p. 89.

[31] Matilda Joslyn Gage, *Woman, Church and State* (1893), Persephone Press, Massachusetts, 1980, p. 98.

[32] Mary Daly, *Gyn/Ecology: The Metaethics of Radical Feminism*, Beacon Press, Boston, 1978.

[33] Kathleen Barry, *Female Sexual Slavery*, 2nd edn, New York University Press, New York & London, 1984, pp. 8, 296.

[34] Christine Delphy, *Close to Home: A Materialist Analysis of Women's Oppression*, translated and edited by Diana Leonard, Hutchinson, London, 1984, p. 180.

[35] Hooks, op. cit., p. 24.

[36] Peggy Seeger, in Rowland, *Women Who Do*, p. 229.

[37] Jersild, in Don E. Hamachek, *Encounters With the Self*, 2nd edn, Holt, Rinehart & Winston, New York, 1978, p. 3.

[38] Inge Broverman, Donald Broverman et al., 'Sex-role Stereotypes and Clinical Judgements of Mental Health', *Journal of Consulting and Clinical Psychology* 34, 1, 1970.

[39] Judith Bardwick and Elizabeth Douvan, 'Ambivalence: The Socialisation of Women', in Judith Bardwick (ed.), *Readings on the Psychology of Women*, Harper & Row, New York, 1972, p. 52.

[40] Kay Deaux, *The Behaviour of Women and Men*, Brooks/Cole, California, 1976.

[41] Daly, op. cit., p. 374.

[42] Matilda Joslyn Gage, cited in Dale Spender, *Women of Ideas and What Men Have Done to Them*, Routledge & Kegan Paul, London and Boston, 1982, p. 239.

[43] Andrea Dworkin, *Right-Wing Women: The Politics of Domesticated Females*, Women's Press, London, 1983, p. 22. This will be expanded in chapter 5.

[44] MacKinnon, op. cit., p. 520.

[45] ibid, p. 530.

[46] Susan Brownmiller, *Femininity*, Fawcett Columbine, New York, 1984, p. 236.

[47] Jill Julius Matthews, *Good and Mad Women: The Historical Construction of Femininity in Twentieth Century Australia*, Allen & Unwin, Sydney, 1984, pp. 7, 8.

[48] Dworkin, op. cit., p. 19.

[49] Hélène Cixous, 'The Laugh of the Medusa', in Elaine Marks and Isabelle de Courtivron (eds), *New French Feminisms: An Anthology*, Harvester, Brighton, 1980, p. 248.

[50] Lou Benson, *Images, Heroes and Self-perceptions: The Struggle for Identity—From Mask-wearing to Authenticity*, Prentice-Hall, New Jersey, 1974, p. 53.

[51] Ellen Dickstein, 'Self and Self-esteem: Theoretical Foundations and their Implications for Research', *Human Development* 20, 1977, p. 139.

[52] Boston Women's Health Collective, *Our Bodies, Ourselves*, Simon & Schuster, New York, 1973. p. 6.

[53] Dworkin, op. cit., p. 21.

[54] William James, 'The Self', in Chad Gordon and Kenneth J. Gergen (eds) *The Self in Social Interaction, vol. 1: Classic and Contemporary Perspectives*, New York, Wiley, 1968, p. 42.

[55] D. Wrong, *Power: Its Forms, Bases and Uses*, Basil Blackwell, Oxford, 1979, p. 111. This is extended in the Conclusion.

[56] John Berger, *Ways of Seeing*, Viking, New York, 1972, pp. 46–7.

[57] MacKinnon, op. cit., p. 539.

[58] Daly, op. cit., p. 355.

[59] Anne Koedt, ' "Politics of the Ego", a Manifesto for New York Radical Feminists', in Anne Koedt, Ellen Levine and Anita Rapone (eds), *Radical Feminism*, Quadrangle/New York Times, New York, 1973, p. 379.

[60] Hartsock, op. cit., p. 60.

[61] Glenys Huws, 'The Conscientization of Women: A Rite of Self-initiation with the Flavour of a Religious Conversion Process', *Women's Studies International Forum* 5, 5, 1982, p. 407.

[62] Erik Erikson, cited in Miriam Dixson, *The Real Matilda. Women and Identity in Australia, 1778–1975*, Penguin, Melbourne, 1976, p. 14.

[63] Dale Spender, *Man Made Language*, Routledge, & Kegan Paul, London, 1980.

[64] See for example Edna Ryan and Anne Conlan, *Gentle Invaders*, Nelson, Melbourne, 1975.

[65] The Australian government's introduction of a $250 fee for tertiary students in 1987, however, definitely discriminates against mature-aged women.

[66] See for example Patricia Niles Middlebrook, *Social Psychology and Modern Life*, 2nd edn, Knopf, New York, 1980.

[67] Paul Schilder, 'The Image and Appearance of the Human Body' cited in Gordon & Gergen, op. cit., p. 107.

[68] Joan Cocks, 'Suspicious Pleasures: On Teaching Feminist Theory', in Margo Culley and Catherine Portuges (eds), *Gendered Subjects: The Dynamics of Feminist Teaching*, Routledge & Kegan Paul, Boston, 1985, p. 178.

[69] See for example Middlebrook, op. cit.

[70] Leghorn and Parker, op. cit., p. 243.

[71] Mary Astell, in Catherine Rogers (ed.), *Before Their Time: Six Women Writers of the 18th Century*, Frederick Ungar, New York, 1979, cited in Spender, op. cit.

[72] Mariana W. Johnson, quoted in Lerner, op. cit., p. 343.

[73] 'The Protection of Girls and Young Women', *Woman's Voice*, May 1905, extract in Kay Daniels and Mary Murnane (eds), *Uphill All The Way: A Documentary History of Women in Australia*, University of Queensland Press, Brisbane, 1980, p. 279.

[74] Clare Burton, *Subordination: Feminism and Social Theory*, Allen & Unwin, Sydney, 1985, p. 133.

[75] Fanny Wright, cited in Spender, op. cit. p. 123.

1 Sex, Biology and Self

[1] Farley Kelly, 'Mrs. Smyth and the Body Politic: Health Reform and Birth Control in Melbourne', in Margaret Bevege, Margaret James and Carmel Shute (eds), *Worth Her Salt: Women at Work in Australia*, Hale & Iremonger, Sydney, 1982, p. 215.

[2] See, for example, Shirley Weitz, *Sex Roles: Biological, Psychological and Social Foundations*, Oxford University Press, New York, 1977; Irene Frieze, Jacquelynne Parsons, Paula Johnson, Diana Ruble, Gail Zellman, *Women and Sex Roles: A Social Psychological Perspective*, Norton, New York, 1978; and Barbara Lusk Forisha, *Sex Roles and Personal Awareness*, General Learning Press, Morristown, New Jersey, 1978.

[3] Ashley Montagu, *The Natural Superiority of Women*, Macmillan, London, 1968.

[4] Valerie Solanis, quoted in Sue Cox (ed.), *Female Psychology: The Emerging Self*, 2nd edn, Science Research Associates, Palo Alto, California, 1976, p. 29.

[5] Adler, quoted in Ann Oakley, *Subject Women*, Pantheon, New York, 1981, p. 43.

[6] Forisha, op. cit.

[7] Alan Alda, in *Ms.* 1975, quoted in Cox, op. cit.

[8] See Alexandra Kaplan and Mary Anne Sedney, *Psychology and Sex Roles: An Androgynous Perspective*, Little, Brown, Boston, 1980.

[9] Both studies cited, ibid., p. 101.

[10] Hilary Lips and Nina Lee Colwill, *The Psychology of Sex Differences*, Prentice-Hall, New Jersey, 1978, p. 207.

[11] Seymour Parker and Hilda Parker, 'The Myth of Male Superiority: Rise and Demise', *American Psychologist* 81, 2, 1979, pp. 289−309.

[12] Joanna Bunker Rohrbaugh, *Woman: Psychology's Puzzle*, Harvester, Brighton, 1980, p. 331.

[13] ibid.; Susan Brownmiller, *Against Our Will: Men, Women and Rape*, Penguin, Harmondsworth and Melbourne, 1976.

[14] Lesley Rogers, 'Biology and Human Behaviour' in Jan Mercer (ed.), *The Other Half: Women in Australian Society*, Penguin, Melbourne, 1975, emphasis added.

[15] Forisha, op. cit., p. 54; see also Frieze et al. and Weitz for further discussion; also Rhoda Unger, *Female and Male: Psychological Perspectives*, Harper & Row, New York, 1979.

[16] Juanita H. Williams, *Psychology of Women: Behaviour in a Biosocial Context*, Norton, New York, 1977.

[17] Harry Harlow, 'Sexual Behaviour in the Rhesus Monkey', in Frank Beach (ed.), *Sex and Behaviour*, Wiley, New York, 1965, p. 242.

[18] Lancaster, discussed in Forisha, op. cit., p. 62.

[19] ibid., p. 63.

[20] Ruth Hubbard,-'The Emperor Doesn't Wear Any Clothes: The Impact of Feminism on Biology', in Dale Spender (ed.), *Men's Studies Modified: The Impact of Feminism on the Academic Disciplines*, Pergamon, New York and London, 1981, p. 214.

[21] Sarah Blaffer Hrdy, *The Woman That Never Evolved*, Harvard University Press, Cambridge, Massachusetts, 1982, p. 83.

[22] Joke Schrijvers, 'Viricentrism and Anthropology', in Gerrit Huizer and Bruce Mannheim (eds), *The Politics of Anthropology: From Colonization and Sexism Toward a View from Below*, Mouton, The Hague, 1979.

[23] See, for example, Diane Bell, *Daughters of the Dreaming*, McPhee Gribble/ Allen & Unwin, Melbourne, 1983; and Diane Bell and Pam Ditton, *Law: The Old and the New*, Aboriginal History for the Central Australian Aboriginal Legal Aid Service, Canberra, 1980; Fay Gale (ed.), *We Are Bosses Ourselves: The Status and Role of Aboriginal Women Today*, Australian Institute of Aboriginal Studies, Canberra, 1983.

[24] Judith Brown, 'Iroquois Women: An Ethnohistorical Note', in Rayna R. Reiter (ed.), *Toward An Anthropology Of Women*, Monthly Review Press, New York, 1975.

[25] Sally Slocum, 'Woman the Gatherer: Male Bias in Anthropology', in Reiter, op. cit., p. 39.

[26] ibid., p. 63.

[27] Agnes Estioko-Griffin and P. Bion Griffin, 'Woman the Hunter: the Agta', in Frances Dahlberg (ed.), *Woman the Gatherer*, Yale University Press, New Haven and London, 1981.

[28] Betty Hiatt, 'Woman the Gatherer', in Fay Gale (ed.), *Women's Role in Aboriginal Society*, 3rd edn, Australian Institute of Aboriginal Studies, Canberra, 1978, p. 4.

[29] Catherine H. Berndt, in Gale, op. cit.

[30] Bell, op. cit., p. 246.

[31] Lois Paul, 'The Mastery of Work and the Mystery of Sex in a Guatemalan Village', in M. Z. Rosaldo and L. Lamphere (eds), *Woman, Culture and Society*, Stanford University Press, Stanford, California, 1974, p. 293.

[32] Parker and Parker, op. cit.

[33] L. Tiger and J. Shaepher, *Women in the Kibbutz*, Harcourt Brace Jovanovich, New York, 1975.

[34] Helen Mayer Hacker, 'Gender Roles from a Cross-cultural Perspective', in L. Duberman, *Gender and Sex in Society*, Praeger, New York, 1975.

[35] Sherry B. Ortner, 'Is Female to Male as Nature Is to Culture?, in Rosaldo and Lamphere, op. cit.

[36] Williams, op. cit., pp. 2, 3.

[37] Andy Metcalf and Martin Humphries, *The Sexuality of Men*, Pluto Press, Sydney and London, 1985, p. 25.

[38] Janet S. Hyde, and B. J. Rosenberg, *Half the Human Experience. The Psychology of Women*, Heath, Lexington, Massachusetts, 1976, p. 17.

[39] Frank Bidart, *The Book of the Body*, Farrar, Straus & Giroux, New York, 1977.

[40] Hyde and Rosenberg, op. cit., p. 17.

[41] See John Bryson, *Evil Angels*, Penguin, Melbourne, 1985.

[42] Williams, op. cit., p. 3.

[43] Weitz, op. cit., p. 169. Jung's collective unconscious for *men* was based on the archetypes of women discussed here, but he did not consider the archetype of man which may have been collected for women. These appear to revolve around men as brutal and destructive.

[44] Paula Weideger, *Menstruation and Menopause: The Physiology and Psychology, the Myth and the Reality*, Penguin, New York, 1977, p. 87.

[45] ibid., p. 96.

[46] ibid., pp. 106–7. See also Carol Tavris and Carole Wade, *Sex Differences in Perspective*, 2nd edn, Harcourt Brace Jovanovich, New York, 1984.

[47] Wolfgang Lederer, *The Fear of Women*, Harcourt Brace Jovanovich, New York, p. 153.

[48] ibid., pp. viii, 282.

[49] Mary Daly, *Gyn/Ecology: The Metaethics of Radical Feminism*, Beacon Press, Boston, 1978.

[50] Barbara Ehrenreich and Diedre English, *For Her Own Good: 150 Years of The Experts' Advice To Women*, Doubleday, New York, 1979.

[51] Bettelheim is discussed in Adrienne Rich, *Of Woman Born: Motherhood as Experience and Institution*, Virago, London, 1982, p. 102.

[52] See for example Robyn Rowland, 'Technology and Motherhood: Reproductive Choice Reconsidered', *Signs: Journal of Women in Culture and Society* 12, 3, 1987, pp. 512–28; 'Women as living laboratories: the new reproductive technologies', in Josefina Figueria-McDonough and Rosemary Sarri (eds), *The Trapped Woman: Catch 22 in Deviance and Control*, Sage, Newbury Park, 1987.

[53] Estelle Ramey, 'Men's Monthly Cycles (they have them too you know)', *The First Ms. Reader*, Warner, New York, 1973, p. 175.

[54] Katharina Dalton, *The Premenstrual Syndrome*, C. C. Thomas, Springfield, Illinois, 1964; Katharina Dalton, 'The Influence of Mother's Menstruation on Her Child', *Proceedings of the Royal Society for Medicine* 59, 1966, p. 1014, cited in Judith Bardwick, *Psychology of Women: A Study of Bio-Cultural Conflicts*, Harper & Row, New York, 1971.

[55] R. Moos, 'The Development of a Menstrual Distress Questionnaire', *Psychosomatic Medicine* 30, 1968, pp. 853–67.

[56] Margi Ripper, interview broadcast on 'The Coming Out Show', ABC radio, 12 December 1982.

[57] Karen Paige, 'The Curse on Women', *Psychology Today* 2, 1975, pp. 37–9.

[58] Ramey, op. cit., p. 176.

[59] Janet Saltzman Chafetz, *Masculine, Feminine or Human? An Overview of the Sociology of Gender Roles*, 2nd edn, Peacock, Illinois, 1978.

[60] Paige, op. cit., p. 39.

[61] Cited in Lucy Twoomey, 'PMS: Profile of a Mystery "Disease"', *Australian*, 22 December 1982, p. 11.

[62] J. F. O'Connor, E. M. Shelley and Lenore O. Stern, 'Behavioural rhythms related to the menstrual cycle', *International Institute for the Study of Human Reproduction*, New York, 1973. p. 7, cited in Weideger, op. cit.

[63] Sophie Laws, Valerie Hey and Andrea Eagan, *Seeing Red: The Politics of Pre-Menstrual Tension*, Hutchinson, London, 1985.

[64] ibid., p. 57.

[65] ibid.—see the article by Sophie Laws, 'Who needs PMT? A feminist

approach to the politics of pre-menstrual tension', for an analysis of Dalton's position and her commercial involvement.

[66] Valerie Hey, 'Getting Away with Murder; PMT and the Press', ibid., p. 78.

[67] Rex Hersey, 'Emotional Cycles in Man', *Mental Science* 77, 1931, p. 169.

[68] Ramey, op. cit., p. 175.

[69] ibid., p. 177.

[70] Margaret Henderson, cited in Cristine Russell, 'Men, It Seems, Have That Time of the Month Too', *National Times*, 20—25 September 1976.

[71] M. McClintock, 'Menstrual Synchrony and Suppression', *Nature* 229, 1971, pp. 244—5.

[72] Ramey, op. cit., p. 181.

[73] Charles Darwin, *The Descent of Man*, 1871, quoted in Brian Easlea, *Science and Sexual Oppression: Patriarchy's Confrontation with Woman and Nature*, Weidenfeld & Nicolson, London, 1981, p. 141.

[74] ibid., p. 144.

[75] Jill Conway, 'Stereotypes of Femininity in a Theory of Sexual Evolution', *Victorian Studies* XIV, September 1970, p. 53, 54.

[76] Carroll Smith-Rosenberg and Charles Rosenberg, 'The Female Animal; Medical and Biological Views of Woman and her Role in Nineteenth-century America', *Journal of American History* LX, 2, 1973, p. 335.

[77] See some readings on women and education during this period, for example, Joan N. Burstyn, 'Education and Sex: The Medical Case Against Higher Education for Women in England, 1870—1900', *Proceedings of the American Philosophical Society* 117, 2, 1973, pp. 79—89.

[78] See Easlea, op. cit., p. 133.

[79] ibid.; Ehrenreich and English, op. cit.

[80] Ben Barker-Benfield, 'Sexual Surgery in Late-Nineteenth-Century America', *International Journal of Health Services* 5, 2, 1975, p. 285.

[81] Cited in Easlea, op. cit., p. 135.

[82] Ehrenreich and English, op. cit., p. 125.

[83] Adrienne Rich, *Of Woman Born*, p. 143.

[84] For example: Marian E. Carter and David N. Joyce, 'Ovarian Carcinoma in a Patient Hyperstimulated by Gonadotrophin Therapy for In Vitro Fertilization: A Case Report', *Journal of In Vitro Fertilization and Embryo Transfer* 4, 2, 1987, pp. 126—7; P. N. Bamford and S. J. Steele, 'Uterine and Ovarian Carcinoma in a Patient Receiving Gonadotrophin Therapy: A Case Report', *British Journal of Obstetrics and Gynaecology* 89, 1982, pp. 962—4.

[85] James Burt, cited in Gena Corea, *The Hidden Malpractice. How American Medicine Mistreats Women*, 2nd edn, Harper & Row, New York, 1985, p. 317.

[86] Kim Chernin, *Womansize*, The Women's Press, London, 1983, p. 8.

[87] Pat Crotty, 'Women and Body Size', *Woman and Social Change*, Deakin University, Geelong, 1985, p. 148.

[88] Jeanne E. Gullahorn, 'Sex Roles and Sexuality', in Elaine Donelson and Jeanne Gullahorn (eds), *Women. A Psychological Perspective*, Wiley, New York, 1977, p. 193.

[89] Laurel Richardson Walum, *The Dynamics of Sex and Gender: A Sociological Perspective*, Rand McNally, Chicago, 1977, p. 74.

[90] John Stevens, 'New Cancer Tests Show Breasts Can Be Saved', *Age*, 16 April 1987.

91 Jill Louw, interviewed by Sarah Boston with her husband Ken, 'No, No Mastectomy', 'Accent' page, *Age*, 16 February 1983.

92 Kathy Kizilos, 'Mastectomy: Should There Be a Choice?', 'Accent' page, *Age*, 11 March 1983. The article does not indicate whether the survey covered Melbourne doctors only or was statewide.

93 Lin Layram, in Bristol Women's Studies Group (eds), *Half the Sky: An Introduction to Women's Studies*, Virago, London, 1979, p. 105.

94 Gullahorn, op. cit., p. 116.

95 Ellen Frankfurt, *Vaginal Politics*, Bantam, New York, 1973, p. ix.

96 Lisa Tuttle, *Encyclopedia of Feminism*, Longman, London, p. 140.

97 The Boston Women's Health Book Collective, *The New Our Bodies, Ourselves*, Penguin, Melbourne, 1985.

98 Walum, op. cit., p. 74.

2 The Rational/Intellectual Self as Expressed in Science

1 Genevieve Lloyd, *The Man of Reason: 'Male' and 'Female' in Western Philosophy*, Methuen, London, 1984, pp. ix, x.

2 Ruth Bleier, *Science and Gender: A Critique of Biology and its Theories on Women*, Pergamon Press, New York, 1984, p. 196.

3 Dale Spender, *Women of Ideas and What Men Have Done to Them*, Routledge & Kegan Paul, London and Boston, 1982, p. 19.

4 Barbara Ehrenreich and Deidre English, *For Her Own Good: 150 Years of the Experts' Advice to Women*, Doubleday, New York, 1979, p. 17.

5 Louise Michele Newman (ed.), *Men's Ideas/Women's Realities. Popular Science, 1870–1915*, Pergamon Press, New York, 1985.

6 Cited in Wendy Faulkner and Eric Arnold (eds), *Smothered by Invention: Technology in Women's Lives*, Pluto Press, London, 1985, p. 27.

7 Barbara Katz Rothman, *The Tentative Pregnancy: Prenatal Diagnosis and the Future of Motherhood*, Penguin, New York, 1987.

8 Brian Easlea, *Science and Sexual Oppression*, Weidenfeld & Nicolson, London, 1981, pp. ix, 7. See also Brian Easlea, *Fathering the Unthinkable: Masculinity, Scientists and the Nuclear Arms Race*, London, Pluto Press, 1983.

9 Kathy Overfield, 'Dirty Fingers, Grime and Slag Heaps: Purity and the Scientific Ethic' in Dale Spender (ed.), *Men's Studies Modified: The Impact of Feminism on the Academic Disciplines*, Pergamon Press, New York and London, 1981, p. 241.

10 Lesley Rogers, "Biology and Human Behaviour', in Jan Mercer (ed.), *The Other Half: Women in Australian Society*, Penguin, Melbourne, 1975; Bleier, op. cit.

11 Sandra Harding, *The Science Question in Feminism*, Open University Press, Milton Keynes, 1986, p. 21.

12 Robin Winkler, 'Psychology as a Social Problem: How Value-free is "Objective Psychology"?', *Australian Psychologist* 8, 1973, pp. 120–7.

13 Ernest Hilgard, Rita L. Atkinson and Richard C. Atkinson, *Introduction to Psychology*, 7th edn, Harcourt Brace Jovanovich, New York, 1979, p. 567; see also Carolyn Wood Sherif, 'What Every Intelligent Person Should Know about Psychology and Women', in Eloise C. Snyder (ed.), *The Study of Women: Enlarging Perspectives of Social Reality*, Harper & Row, New York, 1979.

[14] See Gena Corea, *The Hidden Malpractice: How American Medicine Mistreats Women*, 2nd edn, Harper & Row, New York, 1985; Helen B. Holmes, Betty B. Hoskins and Michael Gross (eds), *The Custom-Made Child? Women-Centred Perspectives*, Humana Press, New Jersey, 1981.

[15] Bleier, op. cit., pp. vii, 197.

[16] Lynne Spender, *Intruders on the Rights of Men: Women's Unpublished Heritage*, Pandora Press/Routledge & Kegan Paul, London and Boston, 1983; Dale Spender, 'The Gate-keepers: A Feminist Critique of Academic Publishing', in Helen Roberts (ed.), *Doing Feminist Research*, Routledge & Kegan Paul, London, 1981.

[17] Alice Rossi, 'Women in Science: Why so Few?', *Science* 148, 1965, pp. 196–202.

[18] Dorothy Zinberg, 'The Past Decade for Women Scientists—Win, Lose or Draw?', *Trends in Biochemical Sciences* 2, 6, 1977, pp. 123–6.

[19] Libby Curran, 'Science Education: Did She Drop Out or Was She Pushed?', in the Brighton Women in Science Group (eds), *Alice Through the Microscope. The Power of Science Over Women's Lives*, Virago, London, 1980.

[20] Gillian Robertson, 'Women in Physics: A Review of Current Thinking', in *The Australian Physicist* September 1975, pp. 122–5.

[21] Bettina Cass, M. Dawson, D. Temple, W. Willis, Ann Winkler, *Why So Few? Women Academics in Australian Universities*, Sydney University Press, Sydney, 1983.

[22] Ione Fett, 'The Future of Women in Australian Medicine', *Medical Journal of Australia* (special supplement) 2, 1976, p. 33.

[23] Cass et al., op. cit., p. 162.

[24] Marie Keir, 'Women in Science', *New Zealand Science Review*, June 1979, pp. 20–5.

[25] Betty Vetter, 'Working Women Scientists and Engineers', *Science* 207, 4 January 1980, pp. 29–34; Zinberg, op. cit.

[26] Rossi, op. cit.

[27] Fett, op. cit., p. 22.

[28] 'Find Pregnancy During Physicians Residency Need Not be a Problem', *Ob Gyn News* 21, 10, 1986, p. 48.

[29] Susan Britton, 'The Rulers of the Earth (and our Health Services) Are Men', *AMA Gazette* 213, 18 January 1979, pp. 14–18.

[30] Curran, op. cit.

[31] Judith Lorber, *Women Physicians. Careers, Status, and Power*, Tavistock, New York and London, 1984, p. 80.

[32] Rupert Hughes, cited in Carol Tavris and Carole Offir, *The Longest War. Sex Differences in Perspective*, Harcourt Brace Jovanovich, New York, 1977, p. 34.

[33] Eleanor Maccoby and Carol Jacklin, *The Psychology of Sex Differences*, Stanford University Press, Stanford, California, 1974.

[34] For example, see Jeanne H. Block in Sue Cox (ed.), *Female Psychology. The Emerging Self*, 2nd edn, St Martin's Press, New York, 1981.

[35] Irene Frieze et al., *Women and Sex Roles: a Social Psychological Perspective*, Norton, New York, 1978.

[36] ibid., p. 64.

[37] Maccoby and Jacklin, op. cit.

[38] See a good coverage of this in Alison Kelly and Helen Weinreich-Haste, 'Science Is for Girls?', *Women's Studies International Quarterly* 2, 3, 1979, pp. 275–93.

[39] Eleanor Maccoby, 'Feminine Intellect and the Demands of Science', *Impact of Science on Society* 20, 1, 1970, pp. 13—28.

[40] Robertson, op. cit.

[41] Janet Sayers, quoting Woolley in Brighton Women in Science Group, op. cit, p. 58.

[42] A. Roe, 'A Psychological Study of Eminent Biologists', *Psychological Monograph* 65, 1951, p. 331, cited in Rossi, op. cit.

[43] Ravenna Helson, 'Women Mathematicians and the Creative Personality', *Journal of Consulting and Clinical Psychology* 36, 2, 1971, pp. 210—20.

[44] Louise Bachtold and Emmy Werner, 'Personality Characteristics of Women Scientists', *Psychological Reports* 31, 1972, pp. 391—6.

[45] ibid., p. 395.

[46] Herbert Walberg, 'Physics, Femininity and Creativity', *Developmental Psychology* 1, 1, 1969, p. 53.

[47] Hilary Lips and Nina Lee Colwill, *The Psychology of Sex Differences*, Prentice-Hall, New Jersey, 1978, p. 175.

[48] J. W. Atkinson, *An Introduction to Motivation*, Van Nostrand, Princeton, New Jersey, 1964; D. C. McClelland, *The Achieving Society*, Van Nostrand, Princeton, New Jersey, 1961.

[49] Matina Horner, 'Fail: Bright Women', *Psychology Today* 3, 6, 1969, pp. 36—8.

[50] Rhoda Unger, *Female and Male: Psychological Perspectives*, Harper & Row, New York, 1979.

[51] Joseph Pleck, 'Male Threat from Female Competence', *Journal of Consulting and Clinical Psychology* 44, 4, 1976, pp. 608—13.

[52] *Australian Women's Studies Newsletter* 1, 1987, p. 12. Compare with Deborah Towns, 'Equality of Opportunity for Women Teachers: Women in the Victorian Teaching Service 1950—1981', in Margaret Bevege, Margaret James and Carmel Shute (eds), *Worth Her Salt. Women at Work in Australia*, Hale & Iremonger, Sydney, 1982.

[53] Gilah Leder, 'Are Our Girls Calculating?', *Educational Magazine* 38, 4. 1981, pp. 2—5.

[54] Judith Firkin, 'Girls, Maths and Science', *Advise* (newsletter of the Advisory Services and Guidance Branch, Victorian Institute of Secondary Education) 26, November 1981, p. 5.

[55] Kelly and Weinreich-Haste, op. cit., p. 280.

[56] Alison Kelly, 'Science for Men Only?', *New Scientist*, 29 August 1974, pp. 538—40.

[57] Kelly and Weinreich-Haste, op. cit.

[58] Val Clarke and Sue Chambers, personal communication on research being carried out at Deakin University.

[59] Kelly, op. cit.

[60] Carolyn Ingvarson, 'Sexism in Science: A Discussion Paper for Conasta 1981', paper delivered while author was with the Equal Opportunity Unit in the Education Department, Victoria, Australia.

[61] Harriet Zuckerman and Jonathan Cole, 'Women in American science', *Minerva* 13, 1, 1975, pp. 82—101.

[62] Barbara Reskin, 'Sex Differentiation and the Social Organisation of Science', *Sociological Inquiry* 48, 1978, p. 9.

[63] 'Science for the People Collective', London, issue no. 29, p. 9.

[64] ibid., p. 10.

[65] Reskin, op. cit. p. 18.

66 Martha White, 'Psychological and Social Barriers to Women in Science', *Science* 170, 1970, pp. 413—16.

67 ibid., p. 414.

68 ibid., p. 413.

69 Lorber, op. cit., p. 14

70 Anne Sayre, *Rosalind Franklin and DNA*, Norton, New York, 1975.

71 ibid., p. 19.

72 Evelleen Richards and Louise Crossley, 'A Woman's Place in Science: Rosalind Franklin and DNA', *Refractory Girl* 16, 1978, pp. 19—26.

73 ibid., p. 23.

74 Ruth Hubbard, 'Reflections on the Story of the Double Helix', *Women's Studies International Quarterly* 2, 3, 1979, pp. 261—74.

75 Richards and Crossley, op. cit., p. 25.

76 Margaret Alic, *Hypatia's Heritage: A History of Women in Science from Antiquity to the Late Nineteenth Century*, Women's Press, London, 1986.

77 Sally Gregory Kohlstedt, 'In from the Periphery: American Women in Science 1830—1880', *Signs: Journal of Women in Culture and Society*, 4, 1, 1978, pp. 81—96.

78 Dale Spender, *Women of Ideas*, p. 171.

79 W. M. O'Neil, 'A Century of Women in the University', *The University of Sydney News* 7 July 1981, p. 130.

80 ibid.

81 Britton, op. cit., p. 17.

82 Kate Campbell, 'A Medical Life', in Patricia Grimshaw and Lynne Strahan (eds), *The Half-Open Door: Sixteen Modern Australian Women Look at Professional Life and Achievement*, Hale & Iremonger, Sydney, 1982, p. 161.

83 ibid., p. 169.

84 See for example Robyn Rowland, 'Women as Living Laboratories: The New Reproductive Technologies' in J. Figueira-McDonough and Rosemary Sarri (eds), *The Trapped Woman: Catch 22 in Deviance and Control*, Sage, Newbury Park, 1987; 'Technology and Motherhood: Reproductive Choice Reconsidered', *Signs: Journal of Women in Culture and Society* 12, 3, 1987, pp. 512—28.

85 Campbell, op. cit., p. 170.

86 See *Wisenet, Journal of the Women in Science Enquiry Network*, 3 November 1985.

87 *Sydney Morning Herald*, 11 November 1972 and *Autralian*, 16 November 1972.

88 Kate Legge, 'Affirmative Action Won't Help: Deans', *Age*, 25 November 1985.

89 Constance Holden, 'NASA Satellite Project: The Boss Is a Woman', *Science* 179, 1973, pp. 48—9.

90 Enid Sichel, quoted in Gloria B. Lubkin, 'Women in Physics', *Physics Today*, April 1971, pp. 23—7.

91 Curran, op. cit., p. 39, 41.

92 ibid., p. 40—1.

93 Wu quoted in Lubkin, op. cit., p. 23.

94 Hubbard, 'Reflections on the Story of the Double Helix', op. cit., p. 273.

95 Ruth Wallsgrove, 'The Masculine Face of Science', in Brighton Women and Science Group, op. cit.

96 Gerry Stimson 'Women in a Doctored World', *New Society* 1 May 1975, pp. 265—7.

97 Jana Thompson, 'Women, Technology and Nuclear Power', *Women's News Service* 20, 1979, p. 20. See Faulkner and Arnold, op. cit., for details of the exclusion of women from technological fields, and the masculinity of technology.

98 Sandra Harding, op. cit., p. 247.

99 Overfield, op. cit., p. 241.

100 Liliane Stehelin, 'Sciences, Women and Ideology', in Hilary Rose and Stephen Rose (eds), *The Radicalization of Science*, Macmillan, London, 1976, p. 87.

101 Anne Fausto-Sterling, 'Women and Science', *Women's Studies International Quarterly* 4, 1, 1981, pp. 41–50.

102 Bleier, op. cit.

3 Culture and Self-creation through Writing

1 Charlotte Perkins Gilman, *The Man-made World or, our Androcentric Culture*, Chalton, New York, 1911, p. 89.

2 Cited in Sandra Gilbert and Susan Gubar, *The Madwoman in the Attic: Re-visions of Nineteenth-century Women Writers*, Yale University Press, New Haven, 1979, p. 3.

3 Audrey T. Rodgers, '"Portrait of a Lady": Images of Women in Twentieth Century American Literature', in Eloise C. Snyder (ed.), *The Study of Women: Enlarging Perspectives of Social Reality*, Harper & Row, New York, 1979, pp. 228–61.

4 Cheri Register, 'American Feminist Literary Criticism: A Bibliographical Introduction', in Josephine Donovan (ed.), *Feminist Literary Criticism. Explorations in Theory*, The University Press of Kentucky, Kentucky, 1975, pp. 1–28.

5 Jessie Bernard, *The Female World*, Free Press, Macmillan, New York, 1981, p. 3.

6 Carol Christ, *Diving Deep and Surfacing*, Beacon Press, Boston, 1980, p. 1.

7 Adrienne Rich, 'When We Dead Awaken: Writing as Re-vision', in *On Lies, Secrets, and Silence: Selected Prose 1966–1978*, Virago, London, 1980, p. 35.

8 Annette Kolodny, 'Dancing Through the Minefield: Some Observations on the Theory, Practice and Politics of a Feminist Literary Criticism', in Dale Spender (ed.), *Men's Studies Modified: The Impact of Feminism on the Academic Disciplines*, Pergamon Press, New York and London, 1981, p. 27.

9 Elaine Showalter, 'Women and the Literary Curriculum', *College English* 32, May 1971, p. 856.

10 Elaine Showalter, 'Women Writers and the Female Experience', in Anne Koedt, Ellen Levine and Anita Rapone (eds), *Radical Feminism*, Quadrangle/New York Times, New York, 1973, p. 394.

11 Judy Turner, 'No Simple Answer', *Australian Book Review*, November 1983, p. 28.

12 P. K. Elkin, 'David Ireland: A Male Metropolis', in Shirley Walker (ed.), *Who Is She? Images of Woman in Australian Fiction*, University of Queensland Press, Brisbane, 1983, pp. 163–73.

13 David Ireland, *City of Women*, Allen Lane, Melbourne, 1981, pp. 154–5.

14 Gilbert and Gubar, op. cit., p. 25.

194 *Woman herself*

15 Delys Bird, 'Towards an Aesthetics of Australian Women's Fiction: *My Brilliant Career* and *The Getting of Wisdom*', *Australian Literary Studies* 2, October 1983, p. 173.
16 Rodney Hall (ed.), *The Collins Book of Australian Poetry*, Collins, Sydney and London, 1981, p. 3.
17 Louisa Lawson, 'The Australian Bush-woman', Boston *Women's Journal*, 27 July 1889, pp. 233–4, reprinted in *Australian Literary Studies* 10, 4, 1982, pp. 500–3.
18 B. J. Zinkhan, 'Louisa Lawson's "The Australian Bush-woman"—A Source for "The Drovers Wife" and "Water them Geraniums"?', *Australian Literary Studies* 10, 4, 1982, pp. 495–9.
19 Frances McInherny, 'Miles Franklin, *My Brilliant Career* and the Female Tradition', *Australian Literary Studies* 9, 3, 1980, p. 281.
20 Gilbert and Gubar, op. cit., p. 76.
21 Carol Burr Megibow, 'The Use of Story in Women's Novels of the Seventies', in Gayle Kimball (ed.), *Women's Culture: The Renaissance of the Seventies*, Metuchen, New Jersey and London, 1981, p. 196.
22 Erica Jong, *At the Edge of the Body*, Holt, Rinehart & Winston, New York, 1979, p. 45.
23 Aldous Huxley, *Words and their Meanings*, Ward Ritchie, Los Angeles, 1940.
24 Gilman, op. cit., p. 89.
25 Cheris Kramarae, 'Proprietors of Language', in Sally McConnell-Ginet, Ruth Borker, Nelly Furman (eds), *Women and Language in Literature and Society*, Praeger, New York, 1980, p. 63.
26 Dale Spender, *Man Made Language*, Routledge & Kegan Paul, London and Boston, 1980.
27 Casey Miller and Kate Swift, *Words and Women: New Language in New Times*, Anchor/Doubleday, New York, 1976.
28 Wendy Martyna, 'The Psychology of the Generic Masculine', in McConnell-Ginet, Borker and Furman, op. cit.,
29 Spender, *Men's Studies Modified*, pp. 6–7.
30 Bernard, op. cit.,
31 Catharine R. Stimpson, 'The Power to Name: Some Reflections on the Avant-grade', in Julia Sherman and Evelyn Beck (eds), *The Prism of Sex: Essays in the Sociology of Knowledge*, University of Wisconsin Press, Wisconsin, 1979, p. 56.
32 Rozsika Parker and Griselda Pollock, *Old Mistresses: Women, Art and Ideology*, Routledge & Kegan Paul, London, 1981, p. 114.
33 Bev Roberts, 'Notes on Women and Literature: The Case Study of Poetry', in Norma Grieve and Patricia Grimshaw (eds), *Australian Women: Feminist Perspectives*, Oxford University Press, Melbourne, 1981, p. 104.
34 Cited in Gilbert and Gubar, op. cit., p. 337.
35 Sheila Rowbotham, *Woman's Consciousness, Man's World*, Penguin, Harmondsworth, 1973, p. 28.
36 Bernard, op. cit., p. 385.
37 Mary Daly, *Gyn/Ecology: The Metaethics of Radical Feminism*, Beacon Press, Boston, 1978, p. 15.
38 Toril Moi, *Sexual/Textual Politics: Feminist Literary Theory*, Methuen, London and New York, 1985.
39 Hélène Cixous, 'The Laugh of the Medusa', in Elaine Marks and Isabelle de Courtivron (eds), *New French Feminisms: An Anthology*, University of Massachusetts Press, Amherst, 1980, pp. 260, 256–7.
40 Elaine Marks, 'Women and Literature in France', *Signs: Journal of Women*

in Culture and Society 3, 4, 1978, p. 837.

41 Carolyn Greenstein Burke, 'Report from Paris: Women's Writing in the Women's Movement', *Signs: Journal of Women in Culture and Society* 3, 4, 1978, p. 852.

42 Ann Rosalind Jones, 'Inscribing Femininity: French Theories of the Feminine', in Gayle Greene and Coppélia Kahn (eds), *Making a Difference: Feminist Literary Criticism*, Methuen, London and New York, 1985, p. 107.

43 Bernard, op. cit., p. 373.

44 See for example the special issue edited by Dale Spender, 'Personal Chronicles, Women's Autobiographical Writings', *Women's Studies International Forum* 10, 1, 1987.

45 Gayle Kimball, 'Women's Culture: Themes and Images', in Kimball, op. cit., p. 22.

46 Rowbotham, op. cit., p. 28.

47 Virginia Woolf, 'Women and Fiction', *Collected Essays 2*, Chatto & Windus, London, 1966, p. 45.

48 Tillie Olsen, *Silences*, Virago, London, 1980, p. 18.

49 Carole Ferrier, 'Introductory Commentary: Women Writers in Australia', in Carole Ferrier (ed.), *Gender, Politics and Fiction. Twentieth Century Australian Women's Novels*, University of Queensland Press, Brisbane, 1985, pp. 15, 16 and *passim.*

50 Tillie Olsen, 'Silences: When Writers Don't Write', in Susan Koppelman Cornillon (ed.), *Images of Women in Fiction: Feminist Perspectives*, Bowling Green University Popular Press, Bowling Green, Ohio, 1972, p. 107.

51 Patricia Meyer Spacks, *The Female Imagination*, Avon Books, New York, 1975, p. 409.

52 Cited in Olsen, *Silences*, p. 18.

53 Woolf, op. cit., p. 45.

54 Tillie Olsen, in Cornillon, op. cit., p. 104.

55 Adrienne Rich, *On Lies, Secrets, and Silences*, p. 43.

56 Gwen Harwood's poem is quoted in Roberts, op. cit.

57 Ellen Moers, *Literary Women*, Women's Press, London, 1978, pp. 4, 6.

58 Gilbert and Gubar, op, cit., p. 17.

59 Cited in Ferrier, op. cit., p. 19.

60 Gilbert and Gubar, op. cit., p. 49.

61 Dale Spender, *Mothers of the Novel*, Pandora Press/Routledge & Kegan Paul, London and New York, 1986.

62 Lynne Spender, *Intruders on the Rights of Men: Women's Unpublished Heritage*, Pandora Press/Routledge & Kegan Paul, London and Boston, 1983.

63 Cited in Elaine Showalter, 'Towards a Feminist Poetics', in Mary Jacobus (ed.), *Women Writing and Writing about Women*, Croom Helm, London, 1979, p. 34.

64 Dale Spender, *Women of Ideas and What Men Have Done to Them*, Routledge & Kegan Paul, London and Boston, 1982.

65 ibid.

66 Lynne Spender, op. cit.

67 Erica Jong, 'Mr. Lowell', *At the Edge of the Body*, Holt, Reinhart & Winston, New York, p. 30.

68 Kay Iseman, 'Barbara Baynton: Woman as "The Chosen Vessel"', *Australian Literary Studies*, 20th anniversary issue, 1983, pp. 25–37.

69 Lucy Frost, 'Barbara Baynton: An Affinity with Pain', in Shirley Walker, *Who Is She?*, p. 65.

70 Register, op. cit., p. 10.

[71] Marcia Land, 'The Silent Woman: Towards a Feminist Criticism', in Arlyn Diamond and Lee R. Edwards (eds), *The Authority of Experience: Essays in Feminist Criticism*, The University of Massachusetts Press, Amherst, 1977.

[72] Mary Ellman, *Thinking about Women*, Macmillan, London, 1968, pp. 32—3.

[73] Stimpson, op. cit., p. 59.

[74] Diamond and Edwards, op. cit., p. xii.

[75] Ellman, op. cit., p. 29.

[76] Moers, op. cit., p. xi.

[77] Spacks, op. cit., p. 406.

[78] Elaine Showalter, *A Literature of Their Own: British Women Novelists from Brontë to Lessing*, Princeton University Press, Princeton, New Jersey, 1977.

[79] Showalter, 'Towards a Feminist Poetics', op. cit., p. 30.

[80] ibid., p. 34.

[81] Gilbert and Gubar, op. cit., p. 78.

[82] See for example Zoë Fairbairns, *Stand We at Last*, Virago, London, 1983.

[83] Stimpson, op. cit., p. 72.

[84] Drusilla Modjeska, *Exiles at Home: Australian Women Writers, 1925—1945*, Angus & Robertson, Sydney, 1981, p. 5.

[85] Jill Roe, 'The Significant Silence: Miles Franklin's Middle Years', *Meanjin* 39, 1, 1980, pp. 48—59.

[86] John Tranter (ed.), *The New Australian Poetry*, Makar Press, Brisbane, 1979.

[87] For an analysis of this poor representation in poetry anthologies see Jennifer Strauss, 'Anthologies and orthodoxies', *Australian Literary Studies* 13, 1, 1987, pp. 87—95.

[88] Hall, op. cit., p. 5.

[89] Judith Brett, 'Where were the Women?', *Meanjin* 42, 3, 1983, p. 368.

[90] Catherine Cuthbert, 'Lesbia Harford and Marie Pitt: Forgotten Poets', *Hecate* 8, 1, 1982, pp. 33—48.

[91] ibid., p. 42.

[92] A full selection of her work is now available in Drusilla Modjeska and Marjorie Pizer (eds), *Poems of Lesbia Harford*, Angus & Robertson, Sydney, 1985.

[93] Roberts, op. cit., p. 101.

[94] Suzanne Juhasz, 'The Critic as Feminist: Reflections on Women's Poetry, Feminism, and the Art of Criticism', *Women's Studies* 5, 1977, pp. 113—27; Elizabeth Janeway, 'Women's Literature', in David Hoffman (ed.), *Harvard Guide to Contemporary American Writing*, Belknap Press, University of Harvard, Cambridge, Massachusetts, 1979.

[95] Alicia Ostriker, 'The Thieves of Language: Women Poets and Revisionist Myth Making', *Signs: Journal of Women in Culture and Society* 8, 1, 1982, pp. 68—90.

[96] Roberts, op. cit., p. 106.

[97] Adrienne Rich, 'When We Dead Awaken', in *Diving into the Wreck: Poems 1971—1972*, Norton, New York, 1973, p. 6.

[98] Adrienne Rich, *On Lies, Secrets, and Silence*, p. 35.

[99] Judith Kegan Gardiner, 'On Female Identity and Writing by Women', *Critical Inquiry* 8, 2, 1981, p. 349.

[100] Cited in Sandra Gilbert, '"My Name Is Darkness": The Poetry of Self-definition', *Contemporary Literature* XVIII, 4, pp. 444, 445.

[101] Kerryn Goldsworthy, 'Review of *Tail Arse Charlie*', *Island Magazine* 16, spring 1983, p. 44.

[102] Annette Stewart, 'The Pursuit of Risk', *Island Magazine* 16, Spring 1983, p. 46.

[103] Alan Laslett, *Opinion*, June 1983.

[104] Graham Burns, *Meanjin* 3, 1984, p. 462.

[105] Adrienne Rich, 'The Stranger', in *Diving into the Wreck*, p. 19.

[106] Gilbert, op. cit., p. 451.

[107] Janeway, op. cit., p. 372.

[108] Rodgers, op. cit.

[109] Rich, op. cit., p. 19.

[110] Adrienne Rich, 'Integrity', in *A Wild Patience has Taken me this Far. Poems 1978–1981*, Norton, New York and London, 1981, p. 8.

[111] Adrienne Rich, *On Lies, Secrets, and Silence*, p. 36.

[112] Denise Levertov, 'The Poet in the World', in J. Webber and J. Grunnan (eds), *Woman as Writer*, Houghton Mifflin, Boston, 1978, p. 85.

[113] Bernard, op cit., p. 435.

[114] See Tillie Olsen, *Silences*, who lists the number of suicides and suspected suicides and they are small, pp. 224, 227.

[115] Anne Sexton, 'Consorting with Angels', in *Live or Die*, Houghton Mifflin, Boston, 1966, p. 20.

[116] Pearl Bell, quoted in Rodgers, op cit., p. 256.

[117] Janeway, op. cit., p. 373.

[118] Juhasz, op. cit., p. 27.

[119] Margaret Atwood, 'Circle/Mud Poems', *You Are Happy*, Harper & Row, New York, 1974, pp. 49–51.

[120] Mona van Duyn, from *To See, to Take* (1970), cited in Ostriker, op. cit., p. 78.

[121] Denise Levertov, 'Song for Ishtar', in *Oh Taste and See*, New Directions, New York, 1962, p. 3.

[122] Janet Montefiori, 'Feminist Identity and the Poetic Tradition', *Feminist Review* 13, February 1983, pp. 69–84; Sydney Janet Kaplan, 'Literary Criticism', Review Essay in *Signs: Journal of Women in Culture and Society* 4, 3, 1979, pp. 514–27.

[123] Ostriker, op. cit., p. 71.

[124] Elizabeth Hardwick, 'The Subjection of Women', in Elaine Showalter (ed.), *Women's Liberation and Literature*, Harcourt Brace Jovanovich, New York, 1971, p. 210.

[125] Alicia Ostriker, 'Body Language: Imagery of the Body in Women's Poetry', in Leonard Michaels and Christopher Ricks (eds), *The State of the Language*, University of California Press, Berkeley, 1980, p. 248.

[126] Marge Piercy, 'Right to Life', in *The Moon Is Always Female*, Knopf, New York, 1980, pp. 95–7.

[127] Juhasz, op. cit., p. 29.

[128] Cixous, op. cit., p. 250.

4   The Labours of Women: Self-definition through Work

[1] *United Nations Commission on the Status of Women*, Newsletter no. 3, 1980; Leghorn and Parker op. cit.

[2] Jessie Bernard, *The Female World*, Free Press, Macmillan, New York, 1981, p. 3.

[3]   Lisa Leghorn and Katherine Parker, *Women's Worth. Sexual Economics and the World of Women*, Routledge & Kegan Paul, Boston, 1981, p. 13.

[4]   Virginia Novarra, *Women's Work, Men's Work: The Ambivalence of Equality*, Martin Boyars: London and Boston, 1980.

[5]   Philanthropic work has an ambiguous status. While ensuring the continuity of caring for those in need, it does enable society to let the work go unrecognized and to avoid employing people to do it.

[6]   Leghorn and Parker, op. cit.

[7]   Bernard, op. cit., p. 519.

[8]   Novarra, op. cit.

[9]   Jill Julius Matthews, 'Deconstructing the masculine universe: The case of women's work', *All Her Labours: Working It Out*, Hale & Iremonger, Sydney, 1984; Rosemary Game and Ann Pringle, *Gender at Work*, Allen & Unwin, Sydney, 1983.

[10]   Matthews, op.cit., pp. 15, 16.

[11]   Novarra, op. cit., p. 19.

[12]   Cited in Leghorn and Parker, op.cit.

[13]   Leghorn and Parker, op. cit., p. 281.

[14]   ibid., pp. 114, 144.

[15]   Novarra, op. cit., pp. 20–1.

[16]   Game and Pringle, op. cit., p. 133.

[17]   Leghorn and Parker, op. cit., p. 3.

[18]   Sheila B. Kamerman, 'Women and Family in Industrialized Societies', *Signs: Journal of Women in Culture and Society* 4, 4, 1979, p. 632.

[19]   Louise Tilly and Joan Scott, *Women, Work and Family*, Holt, Rinehart and Winston, New York, 1978, p. 227.

[20]   Barbara Ehrenreich, *The Hearts of Men: American Dreams and the Flight from Commitment*, Pluto Press, London, 1983, p. 7.

[21]   Tilly and Scott, op. cit., p. 229.

[22]   Heidi I. Hartmann, 'The family as the locus of gender, class and political struggle: The example of housework', *Signs: A Journal of Women in Culture and Society* 6, 3, 1981, pp. 366–94.

[23]   Tilly and Scott, op. cit., p. 232.

[24]   Chris Middleton, 'Patriarchal exploitation and the rise of English capitalism', in Eva Gamarnikow, David Morgan, June Purvis and Daphne, Taylorson (eds), *Gender, Class and Work*, Heinemann, London, 1983, p. 25.

[25]   Christine Delphy, *Close to Home: A Materialist Analysis of Women's Oppression*, translated and edited by Diana Leonard, Hutchinson, London, 1984, p. 18.

[26]   Sheila Rowbotham, *Woman's Consciousness, Man's World*, Penguin, Harmondsworth, 1974, p. 67.

[27]   Michèle Barrett, *Women's Oppression Today: Problems in Marxist Feminist Analysis*, Verso, London, 1980, p. 249.

[28]   Hartmann, op. cit.; Veronica Beechey, 'Women and production: A critical analysis of some sociological theories of women's work' in Annette Kuhn and AnnMarie Wolpe (eds), *Feminism and Materialism*, Routledge & Kegan Paul, London, 1978.

[29]   Game and Pringle, op. cit., pp. 14, 16.

[30]   Ann Game and Rosemary Pringle, 'Beyond Gender at Work: Secretaries', in Norma Grieve and Ailsa Burns (eds), *Australian Women: New Feminist Perspectives*, Oxford University Press, Melbourne, 1986, p. 276.

[31] Game and Pringle, op. cit., p. 22.

[32] Barrett, op. cit.

[33] Hartmann, op. cit.

[34] Mary McIntosh, 'The State and the Oppression of Women', in Kuhn and Wolpe, op. cit., p. 255.

[35] Tilly and Scott, op. cit., p. 230.

[36] Women's Bureau, Department of Employment and Industrial Relations, *Facts on Women at Work in Australia, 1983*, Australian Government Publishing Service, Canberra, 1985, p. 4.

[37] Sandra Eccles, 'Women and the Australian Labour Force', in Dorothy Broom (ed.), *Unfinished Business: Social Justice for Women in Australia*, Allen & Unwin, Sydney, 1984.

[38] Women's Bureau, op. cit., p. 34.

[39] ibid., p. 12.

[40] Myra H. Strober, 'Woman & Man in the World of Work: Present and Future', in L. A. Cater, A. F. Scott and M. Wendy (eds), *Women and Man: Changing Roles*, Praeger, New York, 1977.

[41] Francine D. Blau, 'Women in the Labor Force: An Overview', in Jo Freeman (ed.), *Women. A Feminist Perspective*, 2nd edn, Mayfield, Palo Alto, California, 1979.

[42] Mary Ann O'Loughlin and Bettina Cass, 'Married Women's Employment Status and Family Income Distribution', paper presented at 54th ANZAAS Congress, Canberra, 1984, p. 4.

[43] See Blau, op. cit.; and Kate Purcell, 'Militancy and Acquiescence amongst Women Workers', in S. Burman (ed.), *Fit Work for Women*, Croom Helm, London, 1979.

[44] Eccles, op. cit., p. 88.

[45] Women's Bureau, *Women and Work*, April 1986—figures for November 1985.

[46] Bettina Berch, *The Endless Day: The Political Economy of Women and Work*, Harcourt Brace Jovanovich, New York, 1982.

[47] *Comrades* ABC–TV, 8 December 1986.

[48] Strober, op. cit.

[49] Linda Rubinstein, 'Women, Work and Technological Change', in Elizabeth Windschuttle (ed.), *Women, Class and History: Feminist Perspectives on Australia 1788–1978*, Fontana, Sydney, 1980.

[50] Game and Pringle, op. cit., p. 278.

[51] Eccles, op. cit., p. 85.

[52] Lynn Beaton, 'The Importance of Women's Paid Labour: Women at Work in World War II', in Margaret Bevage, Margaret James and Carmel Shute (eds), *Worth Her Salt: Women at Work in Australia*, Hale & Iremonger, Sydney, 1982, p. 88.

[53] Joan Curlewis, 'Women Working in Heavy Industry in World War II', *All Her Labours: Working It Out*, Hale & Iremonger, Sydney, 1984.

[54] Beaton, op. cit., p. 97.

[55] ibid.

[56] John Bowlby, *Child Care and the Growth of Love*, Penguin, Harmondsworth, 1953.

[57] See for example Claire Etaugh, 'Effect of Maternal Employment on Children: A Review of Recent Research', *Merrill-Palmer Quarterly* 20, 2, 1974, pp. 71–98.

[58] Alexandra Kaplan and Mary Anne Sedney, *Psychology and Sex Roles. An Androgynous Perspective*, Little, Brown, Boston, 1980; G. Russell, *The Changing Role of Fathers*, University of Queensland Press, Brisbane, 1983.

[59] Berch, op. cit.

[60] Juliet Mitchell, in Robyn Rowland (ed.), *Women Who Do and Women Who Don't, Join the Women's Movement*, Routledge & Kegan Paul, London, 1984.

[61] Margaret Simons, 'Poor Relations of the Rag Trade', 'Accent' page, *Age*, 6 April 1983. Subsequently new conditions for outworkers, to operate from 5 November 1987, were approved by the Arbitration Committee. The case on behalf of outworkers was run by the Clothing and Allied Trades Union with Anna Booth as general secretary. Workers' new conditions include improved pay, double time at weekends, superannuation and long-service leave. Employers will be fined $1000 if they breach the award. Yet the issues discussed here still apply to those oppressed people who are unaware of the new conditions or are afraid to face an employer.

[62] Berch, op. cit., and Strober, op.cit.

[63] Australian, Bureau of Statistics survey.

[64] Women's Bureau, *Women and Work*, op. cit.

[65] Frank Jones 'Income Inequality', in Broom, op cit., p. 102.

[66] Bruce Chapman, in Broom, op. cit.

[67] Helen Marchant, 'Union Strategies and the Role of the State in Determining Women's Wages in the 1940s', ANZAAS Congress, Canberra, 1984.

[68] Jennie Bremner, 'In the Cause of Equality: Muriel Heagney and the Position of Women in the Depression', in Bevage, James and Shute, op. cit., p. 293.

[69] Marchant, op. cit., p. 15. See also Ann Curthoys, 'Equal Pay, a Family Wage, or both: Women Workers, Feminists, and Unionists in Australia since 1945', paper to Women's Studies Conference, Sydney, September 1985.

[70] Strober, op. cit.

[71] Chapman, op. cit., p. 118.

[72] Blau, op. cit., p. 284.

[73] Ann Curthoys 'The Sexual Division of Labour: Theoretical Arguments', in Grieve and Burns, op.cit.

[74] Berch, op. cit., p. 8.

[75] Kaplan and Sedney, op. cit.

[76] Peter McDonald, 'Can the Family Survive', *Australian Society* 2, 11, 1 December 1983; O'Loughlin and Cass, op. cit.

[77] Strober, op. cit.

[78] Leghorn and Parker, op. cit., p. 185.

[79] Kaplan and Sedney, op. cit., p. 307.

[80] The Council of Social Service of New South Wales, 'Unemployed Women— A Research Report', *Refractory Girl*, 18–19, 1979, p. 15.

[81] Meredith Edwards, 'Economics of Home Activities', *Australian Journal of Social Issues* 15, 21, 1980 in Broom, op. cit.; and Meredith Edwards, *Financial Arrangements Within Families*, a research report for the National Women's Advisory Council, February 1981.

[82] D. Gillespie, 'Who Has the Power? The Marital Struggle', *Journal of Marriage and the Family* 33, 1971, pp. 445–58.

[83] Curlewis, op. cit.

[84] Hartmann, op. cit.

85 Game and Pringle, op. cit., p. 126.
86 Jan Harper and Lyn Richards, *Mothers and Working Mothers*, Penguin, Melbourne, 1979.
87 Lois Bryson, 'Housewives, Housework and Housepower', presidential address to the Women's Studies section, 53rd ANZAAS Congress, Perth, 1983, pp. 10, 19.
88 Game and Pringle, op. cit., p. 135.
89 Bryson, op. cit., pp. 22, 25.
90 Bernard, op. cit.
91 The term 'housewife' is a cross-class term. In language the concept is created of a woman wedded to—entrapped and within—a house. Isolation, lack of mobility and servicing are implied. The term 'houseworkers' will be used instead.
92 Rowbotham, op. cit., p. 70.
93 Berch, op. cit., p. 94.
94 Game and Pringle, op. cit., p. 124.
95 Ann Oakley, *Women's Work: The Housewife, Past and Present*, Pantheon, New York, 1974, p. 222.
96 Edwards, op. cit.
97 Berch, op. cit.
98 Leghorn and Parker, op.cit., p. 170.
99 Oakley, op. cit., p. 227.
100 Alva Myrdal and Viola Klein, *Women's Two Roles: Home and Work*, Routledge & Kegan Paul, London, 1956, pp. 148–9.
101 Rowbotham, op. cit., p. 74.
102 Kaplan and Sedney, op. cit.
103 Rowbotham, op. cit., p. 75.
104 Leghorn and Parker, op. cit., p. 172.
105 For the experience of RSI, see Robyn Rowland, 'Overuse Injury: the experience from a patient's perspective', *Australian Journal of Physiotherapy* 33, 4, 1987, pp. 262–7.
106 Kerry Liddicoat, 'The Health Implications of Screen-based Equipment for Women Workers', paper given at 54th ANZAAS Congress, Canberra, 1984.
107 Susan Thompson and Jenni Neary, 'Technology, Work and Women's Health', paper given at 53rd ANZAAS Congress, Perth, 1983.
108 Rosalind Petchetsky, 'Workers, Reproductive Hazards and the Politics of Protection: An Introduction', *Feminist Studies* 5, 2, 1979, pp. 233–45.
109 Baiba Irving, 'Women, Work and Health', *Refractory Girl* 18–19, 1979, pp. 23–32.
110 Petchetsky, op. cit., p. 239.
111 Irving, op. cit.
112 Thompson and Neary, op. cit.
113 Rubinstein, op. cit.
114 Women's Bureau, Department of Employment and Industrial Relations, ''Jennie George: First Woman on the ACTU Executive', *Women and Work Bulletin* 5, 3, 1983, p. 1.
115 Women's Bureau, *Facts on Women at Work in Australia 1983*, op. cit.
116 Berch, op. cit.
117 See for example Kate Purcell, 'Militancy and Acquiescence among Women Workers', in Burman, op. cit.

[118] Wendy Brady, "'Serfs of the Sodden Scone"?': Women Workers in the Western Australian Hotel and Catering Industry, 1900–1925', paper given at 53rd ANZAAS Congress, Perth, 1983, p. 20.

[119] Hartmann, op. cit.

[120] Barbara Ehrenreich, 'The Women's Movement: Feminist and Anti-Feminist', *Radical America* 15, 1981, p. 99.

5　Love, Sexuality and Friendship

[1] Janice Raymond, *A Passion for Friends: Toward a Philosophy of Female Affection*, Beacon Press, Boston, 1986, p. 11.

[2] Judith M. Bardwick, *In Transition: How Feminism, Sexual Liberation and the Search for Self-Fulfilment Have Altered America*, Holt, Rinehart & Winston, New York, 1979, p. 126.

[3] Nancy Chodorow, *The Reproduction of Mothering: Psychoanalysis and the Sociology of Gender*, University of California Press, Berkeley, 1978, p. 167.

[4] ibid., p. 169.

[5] See for example Warren Farrell, *The Liberated Man: Beyond Masculinity: Freeing Men and their Relationships with Women*, Ramdom House, New York, 1974; Andrew Tolson, *The Limits of Masculinity*, Tavistock, London, 1977; Elizabeth H. Pleck and Joseph H. Pleck, *The American Man*, Prentice-Hall, Englewood Cliffs, New Jersey, 1980.

[6] Chodorow, op. cit., p. 199.

[7] Nancy Henley and Jo Freeman, 'The Sexual Politics of Inter-personal Behaviour' in Jo Freeman (ed.), *Women: A Feminist Perspective*, 2nd edn, Mayfield, Palo Alto, California, 1979, p. 478.

[8] Vincent Kavalovski, 'Men and the Dream of Brotherhood', in R. Lewis (ed.), *Men in Different Times: Masculinity Today and Tomorrow*, Prentice-Hall, Englewood Cliffs, New Jersey, 1981.

[9] Joseph Pleck, 'Men's Power with Women, Other Men and Society: A Man's Movement Analysis', in Lewis, op. cit., p. 237.

[10] Rowbotham, *Woman's Consciousness, Man's World*, Penguin, Harmondsworth, 1973, p. 62.

[11] Rhoda Unger, *Female and Male: Psychological Perspectives*, Harper & Row, New York, 1979.

[12] Laurel Richardson Walum, *The Dynamics of Sex and Gender: a Sociological Perspective*, Rand McNally, Chicago, 1977, p. 172.

[13] Shulamith Firestone, *The Dialectic of Sex: The Case for Feminist Revolution*, Paladin, London, 1972, p. 133.

[14] Nancy Friday, *My Mother My Self: a Daughter's Search for Identity*, Fontana, New York, 1977, p. 54.

[15] Firestone, op. cit., pp. 121, 122, 123.

[16] Betty Friedan, *The Feminine Mystique*, Penguin, Harmondsworth and Melbourne, 1965, p. 280.

[17] Elaine Donelson and Jeanne Gullahorn (eds), *Women: A Psychological Perspective*, Wiley, New York, 1977, p. 148.

[18] Jane Lazarre, 'Loving Men', *Feminist Studies* 6, 1, 1980, p. 214.

[19] Ann Oakley, *Subject Women*, Pantheon, New York, 1981, pp. 239, 256.

[20] Walum, op. cit., p. 167.

[21] W. R. Gove and J. F. Tudor, 'Adult Sex Roles and Mental Illness', *American Journal of Sociology* 78, 1973, pp. 812–35; Jessie Bernard, *The Future of Marriage*, World, New York, 1972.

[22] Unger, op. cit.

23 See for example Jocelynne Scutt, *Even in the Best of Homes: Violence in the Family*, Penguin, Melbourne, 1983. The Victorian state government is considering changing the law which allows rape within marriage.

24 Women's Policy Co-Ordination Unit, Department of Premier and Cabinet, Victoria, *Criminal Assault in the Home: Social and Legal Responses to Domestic Violence*, 1985, p. 5. This report indicates clearly the under-reporting by women of domestic violence.

25 See for example 'Dividing the Cost of Child-support', *Age*, 31 October 1986, p. 22.

26 Karen Lindsey, 'Madonna or Whore?', *Women's News Service* 25, 1980, p. 5.

27 Charlotte Bunch, 'Not for Lesbians Only', in *Building Feminist Theory. Essays from Quest, a Feminist Quarterly*, Longman, New York, 1981, p. 69.

28 Adrienne Rich, 'Compulsory Heterosexuality and Lesbian Existence', *Signs: Journal of Women in Culture and Society* 5, 4, 1980, p. 641.

29 Miriam Dixson, *The Real Matilda: Women and Identity in Australia 1778–1975*, Penguin, Melbourne, 1976, p. 27.

30 Dale Spender, *Man Made Language*, Routledge & Kegan Paul, London and Boston, 1980.

31 Beatrix Campbell, 'A Feminist Sexual Politics: Now You See It, Now you Don't' in Mary Evans (ed.), *The Woman Question: Readings on the Subordination of Women*, Fontana, Oxford, 1982, p. 131.

32 Laws and Schwartz, *Sexual Scripts: The Social Construction of Female Sexuality*, Dryden Press, Hinsdale, Illinois, 1977, p. 15.

33 Alfred Kinsey, *Sexual Behaviour in the Human Female*, 1953; Havelock Ellis, *Studies in the Psychology of Sex*, 3 vols, F. A. Davis, Philadelphia, 1913, 1927, 1934; William Masters and Virginia Johnson, *Human Sexual Responses*, Churchill, London, 1966.

34 Shere Hite, *The Hite Report: A Nationwide Study of Female Sexuality*, Dell, New York, 1981.

35 See for example Marion Kaplan, 'Prostitution, Morality Crusades and Feminism: German-Jewish Feminists and the Campaign against White Slavery', *Women's Studies International Forum* 5, 6, 1982, pp. 619–28; and Sheila Jeffreys, *The Spinster and her Enemies: Feminism and Sexuality 1880–1930*, Pandora Press/Routledge & Kegan Paul, London, 1985; and Zöe Fairbairns, *Stand We At Last*, Virago, London, 1983.

36 Barbara Cameron, 'The Flappers and the Feminists: A Study of Women's Emancipation in the 1920s', in Margaret Bevage, Margaret James and Carmel Shute (eds), *Worth Her Salt: Women at Work in Australia*, Hale & Iremonger, Sydney, 1982, p. 262.

37 Sheila Jeffreys, ' "Free from All Uninvited Touch of a Man": Women's Campaigns Around Sexuality, 1880–1914', *Women's Studies International Forum* 6, 5, 1982, p. 631.

38 Andrea Dworkin, *Right-Wing Women: The Politics of Domesticated Females*, Women's Press, London, 1983, p. 80. See also *Intercourse*, Free Press, MacMillan, New York, 1978.

39 Campbell, op. cit., p. 129.

40 Catharine MacKinnon, 'Feminism, Marxism, Method and the State: An Agenda for Theory', *Signs: Journal of Women in Culture and Society* 7, 3, 1982, pp. 532, 533.

41 Lal Coveney, Margaret Jackson, Sheila Jeffreys, Leslie Kay and Pat Mahony, *The Sexuality Papers; Male Sexuality and the Social Control of Women*, Hutchinson, London, 1984.

42 Lazarre, op. cit., p. 212.

[43] ibid, p. 217.
[44] Raymond, op. cit., pp. 3, 62.
[45] Margaret Atwood, *The Handmaid's Tale*, Jonathan Cape, London, 1986, p. 143.
[46] Pauline Nestor, *Female Friendships and Communities: Charlotte Brontë, George Eliot, Elizabeth Gaskell*, Clarendon Press, Oxford, 1985.
[47] Jeffreys, op. cit., p. 7.
[48] ibid., p. 89.
[49] Cited in Jeffreys, 'Free from All Uninvited Touch', p. 36.
[50] ibid.
[51] Unger, op. cit., p. 279; Robin Penman and Yvonne Stolk, *Not the Marrying Kind: Single Women in Australia*, Penguin, Melbourne, 1983.
[52] Unger, op. cit.
[53] Penman and Stolk, op. cit., p. 83.
[54] Donelson and Gullahorn, op. cit., p. 243.
[55] Penman and Stolk, op. cit., p. 92.
[56] Tricia Bickerton, 'Women Alone', in Sue Cartledge and Joanna Ryan (eds), *Sex and Love: New Thoughts on Old Contradictions*, Women's Press, London, 1984, p. 164.
[57] Penman and Stolk, op. cit., p. 92.
[58] Robyn Rowland, *Women Who Do and Women Who Don't, Join the Women's Movement*, Routledge & Kegan Paul, London and Boston, 1984, pp. 207, 188.
[59] Ziva Kwitney, 'Living Without Them', *Ms Magazine*, October 1975, p. 71.
[60] Jeffner Allen, 'Motherhood: The Annihilation of Women', in Joyce Trebilcot (ed.), *Mothering: Essays in Feminist Theory*, Rowman & Allanheld, Totowa, New Jersey, 1984.
[61] Eva Feder Kittay, 'Womb Envy: An Explanatory Concept', in Trebilcot, op. cit.
[62] Adrienne Rich, *Of Woman Born: Motherhood as Experience and Institution*, Virago, London, 1982.
[63] Jessie Bernard, *The Future of Marriage*, Bantam, New York, 1973.
[64] Elizabeth Badinter, *The Myth of Motherhood: An Historical View of the Maternal Instinct*, Souvenir Press, London, 1981, p. 327.
[65] Chodorow, op. cit.
[66] Barbara Wishart, 'Motherhood within Patriarchy—A Radical Feminist Perspective', in *All Her Labours: Embroidering the Framework*, Hale & Iremonger, Sydney, 1984, pp. 31, 27.
[67] Sara Ruddick 'Preservative Love and Military Destruction: Some Reflections on Mothering and Peace', in Trebilcot, op. cit., p. 217.
[68] Rich, *Of Woman Born*, p. 245.
[69] ibid., p. 235.
[70] Ruddick, op. cit., p. 221.
[71] See Renate D. Klein, 'Doing It Ourselves: Self-insemination', in Rita Arditti, Renate Duelli Klein and Shelley Minden, *Test-tube Women: What Future for Motherhood?*, Routledge & Kegan Paul, London and Boston, 1984.
[72] Jean Renvoize, *Going Solo: Single Mothers By Choice*, Routledge & Kegan Paul, London, 1985.
[73] Miriam D. Mazor, 'Barren Couples', *Psychology Today* 12, 1979, p. 104.
[74] ibid.; Barbara Eck Manning, *Infertility: A Guide for the Childless Couple*, Prentice-Hall, New Jersey, 1977.
[75] Ellen Peck and J. Senderowitz, *Pro-Natalism: The Myth of Mom and Apple Pie*, Crowell, New York, 1974.

[76] P. H. Jamison, L. R. Franzini, R. M. Caplin, 'Some Assumed Characteristics of Voluntarily Childfree Women and Men', *Psychology of Women Quarterly* 4, 1979, pp. 266–73. See also E. Nason and M. Paloma, 'Voluntarily Childless Couples: The Emergence of a Variant Lifestyle', *Sage Research Papers in Social Sciences*, Studies of Marriage and the Family Series, Sage, California, 1976; J. E. Veevers, *Childless by Choice*, Butterworths, Toronto, 1980, and Robyn Rowland, 'An Exploratory Study of the Childfree Lifestyle', *Australian and New Zealand Journal of Sociology* 18, 1, 1982, pp. 17–30.

[77] L. G. Calhoun and J. W. Selby, Voluntary Childlessness, Involuntary Childlessness, and Having Children: A Study of Social Perceptions', *Family Relations* 29, 1980, pp. 181–3.

[78] Personal communication, 1982

[79] M. Movius, 'Voluntary Childlessness—The Ultimate Liberation', *Family Co-ordinator* 25, 1976, pp. 7–63.

[80] Adrienne Rich, *Of Woman Born*, p. 253.

[81] Jo Sutton and Scarlet Friedman, 'Fatherhood: Bringing It All Back Home', in Scarlet Friedman and Sarah Elizabeth (eds), *On The Problem of Men*, Women's Press, London, 1982, pp. 125, 124.

[82] Scarlet Pollock and Jo Sutton, 'Fathers' Rights, Women's Losses', *Women's Studies International Forum* 8, 6, 1985, p. 598.

[83] Barbara Ehrenreich, *The Hearts of Men: American Dreams and the Flight from Commitment*, Pluto Press, London, 1983.

[84] Letter to the *National Times*, 20–26 May 1983, p. 44.

[85] Barbara Hutton, 'Dividing the Cost of Child Support', *Age*, 31 October 1986, p. 22.

[86] Patrick Heffernan, ibid.

[87] Phyllis Chesler, *Mothers on Trial: The Battle for Children and Custody*, Seal, Washington, 1986, p. xi.

[88] Pollock and Sutton, op. cit., p. 598.

[89] ibid., p. 597.

[90] Badinter, op. cit., p. 324.

[91] Pollock and Sutton, op. cit., p. 598.

[92] Elizabeth Ward, 'Rape of Girl-children by Male Family Members', *Australian and New Zealand Journal of Criminology* 15, 1982, p. 91.

[93] Judith Herman and Lisa Hirschman, 'Father–daughter Incest', in Sue Cox (ed.), *Female Psychology: The Emerging Self*, 2nd edn, op. cit.

[94] ibid.

[95] Elizabeth Ward, *Father–Daughter Rape*, Women's Press, London, 1984, pp. 140–1.

[96] Sydney Women Against Incest, *Breaking The Silence*, a report based upon the findings of the Women Against Incest phone-in survey, Sydney, 1985.

[97] Sylvia Kinder, in Rowland (ed.), *Women Who Do and Women Who Don't*.

[98] Herman and Hirschman, op. cit., p. 213.

[99] Ward, *Father–Daughter Rape*, pp. 164–5.

[100] Sydney Women Against Incest, op. cit. p. 18.

[101] Ward, *Father–Daughter Rape*, p. 195.

[102] Ward, 'Rape of girl-children', p. 91.

[103] Louise Bernikow, *Among Women*, Harmony, New York, 1980 p. 117.

[104] Mary Daly, *Gyn/Ecology: the Metaethics of Radical Feminism*, Beacon Press, Boston, 1978, p. 379.

[105] Raymond, op. cit., p. 4.

[106] Bardwick, *In Transition*, op. cit.

[107] Bernard, *The Female World*, Free Press, Macmillan, New York, 1981, p. 291.
[108] Jessie Bernard, 'Homosociality and Female Depression', *Journal of Social Issues* 32, 4, pp. 230, 231.
[109] C. Safilios-Rothschild, 'Toward a Social Psychology of Relationships, *Psychology of Women Quarterly* 5, 3, 1981, p. 380.
[110] Bernikow, op. cit., p. 115.
[111] Bernard, op. cit., p. 293.
[112] Nestor, op. cit., p. 24.
[113] Jeffreys, *The Spinster and Her Enemies*, op. cit.
[114] Bernard, op. cit., p. 98.
[115] Carroll Smith-Rosenberg, 'The Female World of Love and Ritual: Relations between Women in Nineteenth Century America', *Signs: Journal of Women in Culture and Society* 1, 1, 1975, pp. 9—10.
[116] C. Smith-Rosenberg, in E. Dubois, M. J. Buhle, T, Kaplan, G. Lerner and C. Smith-Rosenberg, 'Politics and Culture in Women's History:. Symposium', *Feminist Studies* 6, 1, 1980, p. 63.
[117] Smith-Rosenberg, 'The Female World', p. 14.
[118] ibid., p. 27.
[119] Smith-Rosenberg, 'Politics and Culture', p. 63.
[120] Raymond, op. cit., p. 227.
[121] ibid., pp. 9, 23.
[122] R. Rosenberg, *Beyond Separate Spheres: Intellectual Roots of Modern Feminism*, Yale University Press, Newhaven, 1982, p. 141.
[123] Bernikow, op. cit., p. 77.
[124] Dale Spender and Lynne Spender, *Scribbling Sisters*, Hale & Iremonger, Sydney, 1984.
[125] Toni A. H. McNaron (ed.), *The Sister Bond: A Feminist View of a Timeless Connection*, Pergamon, Oxford, 1985, p. 5.
[126] Jeffreys, *The Spinster and Her Enemies*, op. cit., p. 114.
[127] Judith C. Brown. *Immodest Acts: The Life of a Lesbian Nun in Renaissance Italy*, Oxford University Press, New York, 1986, p. 8.
[128] Rosemary Curb and Nancy Manahan (eds), *Lesbian Nuns: Breaking Silence*, Bantam, New York, 1985.
[129] Brown, op. cit., p. 11.
[130] Jeffreys, *The Spinster and Her Enemies*, p. 109.
[131] Lillian Faderman, *Surpassing the Love of Men: Romantic Friendship and Love Between Women From the Renaissance to the Present*, Morrow, New York, 1981, p. 16.
[132] Sonja Ruehl, 'Sexual Theory and Practice: Another Double Standard', in Sue Cartledge and Joanna Ryan (eds), *Sex and Love: New Thoughts on Old Contradictions*, Women's Press, London, 1983, p. 217.
[133] Charlotte Wolff, *Love Between Women*, Gerald Duckworth, London, 1971, p. 62.
[134] Raymond, op. cit., p. 134.
[135] Rich, *The Dream of a Common Language*, Norton, New York, 1978, pp. 75—6.
[136] Rich, 'Compulsory Heterosexuality', pp. 648—9.
[137] Raymond, op. cit., p. 18.
[138] Haunani-Kay Trask, *Eros and Power: The Promise of Feminist Theory*, University of Pennsylvania Press, Philadelphia, 1986.
[139] Bernikow, op. cit., p. 185.
[140] Trask, op. cit., p. 103.

[141] ibid., p. 115.

[142] Wolff, op. cit., p. 165.

[143] Susan Krieger, *The Mirror Dance*, Temple University Press, 1983.

[144] Nancy Friday, *My Mother My Self: A Daughter's Search For Identity*, Fontana, New York, 1977, pp. 173–5.

[145] Janice G. Raymond, 'From Liberation to Libetarianism: Depoliticising in Drag', paper delivered at the Five College Women Studies Conference on Feminism, Sexuality and Power, 27 October 1986, Mt Holyoke College, p. 9.

Conclusion: Power and Resistance

[1] Jean Lipman-Blumen, *Gender Roles and Power*, Prentice-Hall, New Jersey, 1984, p. 5.

[2] D. H. Wrong, *Power: Its Forms, Basis and Uses*, Basil Blackwell, Oxford, 1979, p. 2.

[3] Chad Gordon and Kenneth J. Gergen (eds), *The Self in Social Interaction, vol. 1: Classic and Contemporary Perspectives*, Wiley, New York, 1968.

[4] Nancy C. M. Hartsock, *Money, Sex and Power: Toward a Feminist Historical Materialism*, North Eastern University Press, Boston, 1985.

[5] Berenice A. Carroll, 'Peace Research: The Cult of power', *Journal of Conflict Resolution* 16, 4, 1972, pp. 586–616.

[6] Elizabeth Janeway, *Powers of the Weak*, Knopf, New York, 1980, p. 11.

[7] Robin Morgan, *The Anatomy of Freedom: Feminism, Physics, and Global Politics*, Anchor Books/Doubleday, Garden City, New York, 1984, p. 200.

[8] Barbara B. Watson, 'On Power and the Literary Text', in Mary Evans (ed.), *The Woman Question: Readings on the Subordination of Women*, Fontana, Oxford, 1982, p. 403.

[9] Carroll, op. cit., p. 607.

[10] Charlotte Bunch, 'Woman Power: The Courage to Lead, the Strength to Follow, and the Sense to Know the Difference', *Ms Magazine*, July 1980, p. 48.

[11] Helene Moglen, 'Power and Empowerment', *Women's Studies International Forum* 6, 2, 1983, p. 131.

[12] Hartsock, op. cit.

[13] Lisa Leghorn and Katherine Parker, *Woman's Worth: Sexual Economics and the World of Women*, Routledge & Kegan Paul, Boston, 1981.

[14] ibid., p. 91

[15] Lipman-Blumen, op. cit., p. 6.

[16] Cited in Anne Summers, *Damned Whores and God's Police*, Penguin, Melbourne, 1975, p. 352.

[17] Kate Millett, *Sexual Politics*, Abacus, London, 1972, p. 43.

[18] Lipman-Blumen, op. cit., p. 102.

[19] ibid., p. 104.

[20] Lisa Stamm and Carol D. Ryff, 'Introduction: An Interdisciplinary Perspective on Women's Power and Influence', in Stamm and Ryff (eds), *Social Power and Influence of Women*, West View Crest for the American Association for the Advancement of Science, 1984.

[21] A number of new books are emerging on feminism within the church itself. See for example Margaret Ann Franklyn (ed.), *The Force of the Feminine: Women, Men and the Church*, Allen & Unwin, Sydney, 1986.

[22] Margaret Stacey and Marion Price, *Women, Power and Politics*, Tavistock, London, 1981.

[23] Vikky Randal, *Women and Politics*, St Martin's Press, New York, 1982.

[24] Cynthia Fuchs Epstein, 'Women and Elites: A Cross-national Perspective', in Cynthia Fuchs Epstein and Rose Laub Coser (eds), *Access to Power: Cross-National Studies of Women and Elites*, Allen & Unwin, London, 1981, p. 12.

[25] Margherita Rendel (ed.), *Women, Power and Political Systems*, Croom Helm, London, 1981, p. 15.

[26] Lipman-Blumen, op. cit., p. 6.

[27] ibid., p. 18.

[28] ibid., p. 181.

[29] ibid., p. 182.

[30] ibid., p. 191.

[31] ibid., p. 192.

[32] Bell Hooks, *Feminist Theory: From Margin to Center*, South End Press, Boston, 1984, p. 121.

[33] Brownmiller, op. cit., p. 103.

[34] Australian society began with sexual violence. The landing of the First Fleet in 1788 was characterized by a two-day debauch described by the surgeon of the *Lady Penrhyn:* 'The convicts got to them [the women] very soon after their landing and the scene of debauchery and riot that ensued during the night might be better concealed than expressed'. Cited in Ruth Teale (ed.), *Colonial Eve: Sources on Women in Australia 1788-1914*, Oxford University Press, Melbourne, 1978, p. 10. See also Summers, op. cit., p. 269.

[35] E. J. Hodgens et al., 'The Offence of Rape in Victoria', *Australian and New Zealand Journal of Criminology* 5, 4, 1972, pp. 225–40.

[36] Ross Barber, 'An Investigation into Rape and Attempted Rape Cases in Queensland', *Australian and New Zealand Journal of Criminology* 6, 4, 1973, pp. 214–30.

[37] *Australian*, 10 April 1982, p. 4.

[38] Daly, op. cit., p. 29.

[39] Brownmiller, *Against Our Will: Men, Women and Rape*, Penguin, Harmondsworth and, Melbourne, 1976, p. 15. Statistics constantly reinforce the regularity of rape. For example in the United States of America it is estimated that a rape occurs every two minutes and one out of every three women in the United States will be sexually assaulted in her lifetime. Alison M. Jaggar, *Feminist Politics and Human Nature*, Harvester Press, Sussex, 1983, p. 94.

[40] There is an extensive emerging literature on feminism and pornography. Some references worth following up are: Andrea Dworkin, *Men Possessing Women*, Putnam-Perigree, New York, 1981; Susan Griffin, *Pornography and Science: Culture's Revenge against Nature*, Women's Press, London, 1981; Laura Lederer (ed.), *Take Back the Night: Women on Pornography*, Morrow, New York, 1980; Robin R. Linden, Darlene R. Pagano, Diana E. H. Russel, Susan Leigh Star (eds), *Against Sadomasochism: A Radical Feminist Analysis*, Frog In The Well, California, 1982; Linda ('Lovelace'), Marciano *Ordeal: An Autobiography*, Citadel Press, 1980; Dusty Rhodes and Sandra McNeill (eds), *Women Against Violence Against Women*, Only Women Press, London, 1985.

[41] Hooks, op. cit., p. 124.

[42] Lipman-Blumen, op. cit., p. 96.

[43] ibid., p. 23.

44 Stamm and Ryff, op. cit., p. 4.
45 Nancy Henley, 'Power, Sex and Non-Verbal Communication', *Berkeley Journal of Sociology* 18, 1973, p. 1.
46 See Marilyn Frye, *The Politics of Reality: Essays in Feminist Theory*, Crossing Press, Trumansburg, New York, 1983.
47 Lipman-Blumen, op. cit., p. 131.
48 Hartsock, op. cit., p. 245.
49 For an outline of some of these technologies see Robyn Rowland, 'Women as Living Laboratories: The New Reproductive Technologies', in Josefina Figueira McDonough and Rosemary Sarri (eds), *The Trapped Woman: Catch 22 in Deviance and Control*, Sage, Newbury Park, 1987; Gena Corea, *The Mother Machine: From Artificial Insemination to Artificial Wombs*, Harper & Row, New York, 1985.
50 See for example N. W. Cohen and L. J. Estner, *Silent Knife—Caesarian Prevention and Vaginal Birth After Caesarian*, Bergin and Garvey, Massachusetts, 1983; and Margaret Pryke, Liselotte Mulhen and Kennett Wade, 'Childbirth and Surgery: The Experience of an Australian Sample of 120 Women During First Birth', *New Doctor* 39, 1986, pp. 22–4.
51 Barbara Seaman and Gideon Seaman, *Women and the Crisis in Sex Hormones*, Rawson, New York, 1977.
52 Susheil Muasher, Jairo Garcia and Howard Jones, 'Experience with Diethylstilbestrol-exposed Infertile Women in a Program of In Vitro Fertilisation', *Fertility and Sterility* 42, 1, 1984, p. 22.
53 Anita Direcks and Helen Bequaert Holmes, 'Miracle Drug, Miracle Baby', *New Scientist*, 6 November 1986, pp. 53–5.
54 Diana Sculley, *Men Who Control Women's Health: The Miseducation of Obstetricians/Gynaecologists*, Houghton Mifflin, Boston, 1980.
55 Muasher *et al.*, op. cit.
56 Catherine Martin, 'A New and Fertile Field for Investment', *Bulletin*, 24 June 1986.
57 Mary O'Brien, *The Politics of Reproduction*, Routledge & Kegan Paul, New York, 1981.
58 While the Victorian government claimed to be concerned about the development of IVF Australia, its Victorian economic development corporation is one of the big institutional shareholders in CP Ventures, whose major single investment so far is $2.6 million into the IVF Australian Trust. See Philip McIntosh, 'Kennan to study IVF sale plans', *Age*, 1 April 1985; and Peter Schumpster, 'Why we are in debt to entrepreneurs', *Age*, 7 April 1986.
59 Ann Oakley, *The Captured Womb: A History of the Medical Care of Pregnant Women*, Basil Blackwell, Oxford, 1984, p. 254.
60 See for example Martin, op. cit.; 'IVF team in share issue bid', *Sun*, 26 June 1986.
61 Barbara Eck-Menning, *Infertility: A Guide for the Childless Couple*, Prentice-Hall, New Jersey, 1977.
62 Miriam D. Mazor, 'Barren Couples', *Psychology Today* 12, 1979, pp. 101–12.
63 See for example Gena Corea, *The Mother Machine*, op. cit.; Rita Arditti, Renate Duelli Klein and Shelley Minden (eds), *Test-Tube Women: What Future for Motherhood?*, Routledge & Kegan Paul, London and Boston, 1984; Gena Corea et al., *Man-Made Woman: How New Reproductive Technologies Affect Women*, Hutchinson, London, 1985.
64 Ken Mao and Carl Wood, 'Barriers to Treatment of Infertility by In

Vitro Fertilisation and Embryo Transfer', *Medical Journal of Australia*, 22 April 1984, p. 532.

65 Renate Klein, 'The Crucial Role of In Vitro Fertilisation in the Social Control of Women', paper delivered to the Women's Hearings on Genetic Engineering and Reproductive Technologies at the European Parliament, 6—7 March 1986, Brussels, p. 7.

66 Robyn Rowland, 'Making Women Visible in the Embryo Experimentation Debate', *Bioethics* 1, 2, 1987, pp. 179—88.

67 Gabor Kovacs et al., 'Induction of Ovulation with Human Pituitary Gonadotrophins', *Medical Journal of Australia*, 12 May 1984, pp. 575—9.

68 A. Birkenfeld et al., 'Effect of Clomiphene on the Uterine and Ovaductal Mucosa', *Journal of In Vitro Fertilisation and Embryo Transfer* 1, 2, 1984, p. 99.

69 B. Henriet, L. Henriet, D. Holhoven and V. Seynave, 'The Letal Effect of Super-ovulation on the Embryo', *Journal of In Vitro Fertilisation and Embryo Transfer* 1, 2, 1984.

70 Direcks and Bequaert Holmes, op. cit.

71 Marian Carter and David Joyce, 'Ovarian Carcinoma in a Patient Hyperstimulated by Gonadotrophin Therapy for In Vitro Fertilization: A Case Study', *Journal of In Vitro Fertilisation and Embryo Transfer* 4, 2, 1987, p. 127.

72 For example: M. Atlas and J. Merczer, 'Massive Hyperstimulation and Borderline Carcinoma of the Ovary: A Possible Association', *Ackta Obstet. Gynaecol. Scand.*, 61, 1982, pp. 261—3.

73 Cited in Françoise Laborie, 'New Reproductive Technologies: News from France and Elsewhere', paper delivered at *Third International Interdisciplinary Congress on Women, Woman's Worlds*, Dublin, July, 1987, p. 15. Forthcoming in *Reproductive and Genetic Engineering: International Feminist Analysis* 1, 1, 1988.

74 Andrew Veitch, 'Human Embryo Research "Should be Forbidden"', *Guardian*, 28 September 1984.

75 'Embryo Centre Fails Licensing Test', *New Scientist* 110, 1505, 24 April 1986, p. 24.

76 See Robyn Rowland, *Surrogate Motherhood: Who Pays the Price?*, unpublished; John Robertson, 'Surrogate Mothers: Not So Novel After All', *Hastings Centre Report* 13, 5, 1983, p. 452; section on surrogacy in *Reproductive and Genetic Engineering: International Feminist Analysis* 1, 1, 1988, forthcoming.

77 Perinatal Statistics Unit of the University of Sydney, *IVF and GIFT Pregnancies, Australia and New Zealand, 1986*. This gives data on 2503 pregnancies in sixteen centres, until September 1984. The number of women who enter programmes but who the technology *fails* to assist in becoming pregnant is very difficult to assess because hospitals do not keep figures on 'failures'. It would make their 'success' rate *very* low. But in the *Interim Report of the In Vitro Fertilization Ethics Committee of Western Australia* of August 1984, results from Dr John Yovich show that until 30 June 1984 only 4 per cent of women ended up with a baby: there was a 10 per cent pregnancy rate. In Britain, for all 30 centres, only 7 per cent of women take home a baby: See *The Second Report of the Voluntary Licensing Authority for Human in Vitro Fertilization and Embryology*, 1987.

78 This quote taken from a paper by Barbara Burton, 'Contentious issues of infertility therapy—a consumer's view', given at the Australian Family Planning Association annual conference, Lorne, March 1985.

79 ibid.

[80] Personal communication.

[81] Personal communication.

[82] Burton, op. cit.

[83] Personal communication from a patient involved in CONCERN, a support group for the infertile in Western Australia.

[84] Dick Teresi and Kathleen McAuliffe, 'Male Pregnancy', *Omni*, December 1985, pp. 51—118. Since this article was published a number of newspapers and journals have carried similar stories in Australia.

[85] ibid.

[86] See 'Transsexuals see IVF Programs as their Chance to Become Mothers', *Sydney Morning Herald*, 7 May 1984.

[87] 'The Man Who Became a Woman', *New Idea*, 22 March 1986.

[88] ibid.

[89] Janice Raymond, *The Transsexual Empire: The Making of the She-Male*, Beacon Press, Boston, 1979. [90] ibid. p. xvii.

[91] Elizabeth Janeway, *Powers of the Weak*, Knopf, New York, 1980.

[92] Hooks, op. cit., p. 90.

[93] Janeway, op. cit., p. 5.

[94] Janeway, op. cit., p. 85.

[95] ibid. p. 14.

[96] Lipman-Blumen, op. cit., p. 9.

[97] Wrong, op. cit., p. 111.

[98] Janeway, op. cit., p. 167.

[99] Lipman-Blumen, op. cit.

[100] ibid., p. 8.

[101] Henley, op. cit., and Nancy Henley, *Body Politics: Power, Sex and Non-Verbal Communication*, Prentice-Hall, New Jersey, 1977.

[102] Lipman-Blumen, op. cit., p. 31.

[103] Andrea Dworkin, *Right-Wing Women: The Politics of Domesticated Females*, Women's Press, London, 1983, p. 234.

[104] Janeway, op. cit., p. 292.

[105] Summers, op. cit., p. 20.

[106] Karen Horney, quoted in Sue Cox (ed.), *Female Psychology: Emerging Self*, 2nd edn, St Martin's Press, New York, 1976, p. 191.

[107] Gerda Lerner, 'Placing Women in History: A 1975 Perspective' in Bernice Carroll (ed.), *Liberating Women's History, Theoretical and Critical Essays*, University of Illinois Press, Chicago, 1976, p. 359.

[108] Janey Stone, 'Brazen Hussies and God's Police: Feminist Historiography and the Depression', *Hecate* 8, 1, 1982, p. 21.

[109] Marie Louise Janssen-Jurreit, *Sexism: The Male Monopoly on History and Thought*, Pluto Press, 1982, translated by Verne Moberg, p. 15.

[110] ibid., p. 16.

[111] Sheila Ryan Johansson, '"Her Story", as History: A New Field or Another Fad', in Bernice Carroll, op. cit., p. 427.

[112] Janssen Jurreit, op. cit., p. 33.

[113] Dale Spender, *Women of Ideas and What Men Have Done to Them*, Routledge & Kegan Paul, London and Boston, 1982, p. 11.

[114] Sophie Watson (ed.), Australian Feminist Interventions, Verso, London, forth-coming.

[115] Hester Eisenstein, 'Feminist Judo: Throwing with the Weight of the State', *Australian Left Review* 96, 1986, p. 22.

[116] Sue Wills, 'Big Visions and Bureaucratic Straightjackets, *Australian Left Review* 96, 1986, pp. 23—4.

[117] Jocelynne Scutt, 'United or Divided? Women "Inside" and Women "Outside" Against Male Law-makers in Australia', *Women's Studies International Forum* 8, 1, 1985, p. 21.

[118] Hester Eisenstein, 'The Gender of Bureaucracy: Reflections on Feminism and the State', in Jacqueline Goodnow and Carol Pateman (eds), *Women, Social Science and Public Policy*, Allen & Unwin, Sydney, 1985, p. 115.

[119] Hooks, op. cit., p. 84.

[120] ibid.

[121] Joan Rothschild, 'Taking our Future Seriously', *Quest: a Feminist Quarterly* 2, 3, 1976, pp. 26, 28, 29.

[122] Charlotte Bunch, 'The Reform Tool Kit', in *Building Feminist Theory: Essays from Quest*, Longman, New York, 1981, p. 196.

[123] Robyn Rowland and Renate Klein, 'Radical Feminism', in Sneja Gunew (ed.), Feminist Knowledge as Critique and Construct, Methuen, London, forthcoming.

[124] Robin Morgan, *Sisterhood Is Global*, Anchor/Doubleday, Garden City, New York, 1984.

[125] Hester Eisenstein, 'The Case for Feminist Optimism', *Octagon Lecture*, Vice-Chancellor's lecture series, University of Western Australia, Perth, 10 June 1986, pp. 2, 1.

[126] Hooks, op. cit., p. 86.

[127] Bunch, 'The Reform Tool Kit', op. cit., p. 194.

[128] Janeway, op. cit., p. 167.

[129] Janeway, op. cit., p. 171.

[130] Nancy Hartsock, 'Political Change: Two Perspectives on Power', *Quest: A Feminist Quarterly* 1, 1, 1974, p. 23.

[131] ibid., p. 13.

Select Bibliography

Alic, Margaret, *Hypatia's Heritage: A History of Women in Science from Antiquity to the late Nineteenth Century*, Women's Press, London, 1986.

All Her Labours: Working It Out, Hale & Iremonger, Sydney, 1984.

Arditti, Rita, Duelli Klein, Renate and Minden, Shelley, *Test-tube Women: What Future for Motherhood?*, Routledge & Kegan Paul, London and Boston, 1984.

Badinter, Elizabeth, *The Myth of Motherhood: An Historical View of the Maternal Instinct*, Souvenir Press, London, 1981.

Bardwick, Judith M., *In Transition: How Feminism, Sexual Liberation and the Search for Self-Fulfilment Have Altered America*, Holt, Rinehart & Winston, New York, 1979.

Barrett, Michèle, *Women's Oppression Today: Problems in Marxist Feminist Thought*, Verso, London, 1980.

Barry, Kathleen, *Female Sexual Slavery*, 2nd edn, New York University Press, New York & London, 1984.

Bell, Diane, *Daughters of the Dreaming*, McPhee Gribble/Allen & Unwin, Melbourne, 1983.

Bell, Diane, and Ditton, Pam, *Law: The Old and the New*, Aboriginal History for the Central Australian Aboriginal Legal Aid Service, Canberra, 1980.

Berch, Bettina, *The Endless Day: The Political Economy of Women and Work*, Harcourt Brace Jovanovich, New York, 1982.

Bernard, Jessie, *The Female World*, Free Press, Macmillan, New York, 1981.

Bernikow, Louise, *Among Women*, Harmony Books, New York, 1980.

Bevage, Margaret, James, Margaret and Shute, Carmel (eds), *Worth Her Salt: Women At Work in Australia*, Hale & Iremonger, Sydney, 1982.

Bleier, Ruth. *Science and Gender: A Critique of Biology and its Theories on Women*, Pergamon Press, London, 1984.

Brown, Judith C. *Immodest Acts: The Life of a Lesbian Nun in Renaissance Italy*, Oxford University Press, New York, 1986.

Brownmiller, Susan, *Femininity*, Fawcett Columbine, New York, 1984.

Brownmiller, Susan, *Against Our Will: Men, Women and Rape*, Penguin, Harmondsworth and Melbourne, 1976.

Building Feminist Theory: Essays from Quest, a Feminist Quarterly, Longman, New York, 1981.

Burton, Clare, *Subordination: Feminism and Social Theory*, Allen & Unwin, Sydney, 1985.

Carroll, Berenice (ed.), *Liberating Women's History, Theoretical and Critical Essays*, University of Illinois Press, Chicago, 1976.

Cartledge, Sue, and Ryan, Joanna (eds), *Sex and Love: New Thoughts on Old Contradictions*, Women's Press, London, 1984.

Cass, Bettina, Dawson, M, Temple, D, Willis, W. and Winkler, Ann, *Why So Few? Women Academics in Australian Universities*, Sydney University Press, Sydney, 1983.

Chesler, Phyllis, *Mothers on Trial: The Battle for Children and Custody*, Seal, Washington, 1986.

Chodorow, Nancy, *The Reproduction of Mothering: Psychoanalysis and the Sociology of Gender*, University of California Press, Berkeley, 1978.

Christ, Carol, *Diving Deep and Surfacing*, Beacon Press, Boston, 1980.

Cohen, N. W., and Estner, L. J., *Silent Knife—Caesarian Prevention and Vaginal Birth After Caesarian*, Bergin and Garvey, Massachusetts, 1983.

Coote, Anna, and Campbell, Beatrix, *Sweet Freedom: The Struggle For Women's Liberation*, Picador, London, 1982.

Corea, Gena, *The Hidden Malpractice: How American Medicine Mistreats Women*, 2nd edn, Harper & Row, New York, 1985.

Cornillon, Susan Koppelman (ed.), *Images of Women in Fiction: Feminist Perspectives*, Bowling Green University Popular Press, Bowling Green, Ohio, 1972.

Coveney, Lal, Jackson, Margaret, Jeffreys, Sheila, Kay, Leslie and Mahony, Pat, *The Sexuality Papers: Male Sexuality and the Social Control of Women*, Hutchinson, London, 1984.

Cox, Sue (ed.), *Female Psychology: The Emerging Self*, 2nd edn, Science Research Associates, Palo Alto, California, 1976.

Culley, Margo, and Portuges, Catherine (eds), *Gendered Subjects: The Dynamics of Feminist Teaching*, Routledge & Kegan Paul, Boston, 1985.

Curb, Rosemary, and Manahan, Nancy (eds), *Lesbian Nuns: Breaking Silence*, Bantam, New York, 1985.

Daly, Mary, *Gyn/Ecology: The Metaethics of Radical Feminism*, Beacon Press, Boston, 1978.

Daniels, Kay and Murnane, Mary (eds), *Uphill All The Way: A Documentary History of Women in Australia*, University of Queensland Press, Brisbane, 1980.

Deaux, Kay, *The Behavior of Women and Men*, Brooks/Cole, California, 1976.

Delphy, Christine, *Close to Home: A Materialist Analysis of Women's Oppression*, translated and edited by Diana Leonard, Hutchinson, London, 1984.

Diamond, Arlyn, and Edwards, Lee R. (eds), *The Authority of Experience: Essays in Feminist Criticism*, University of Massachusetts Press, Amherst, 1977.

Direcks, Anita and Holmes, Helen Bequaert, 'Miracle Drug, Miracle Baby', *New Scientist*, 6 November 1986, pp. 53–5.

Dixson, Miriam, *The Real Matilda. Women and Identity in Australia 1778–1975*, Penguin, Melbourne, 1976.

Donovan, Josephine (ed.), *Feminist Literary Criticism. Explorations in Theory*, University Press of Kentucky, Kentucky, 1975.

Dworkin, Andrea, *Men Possessing Women*, Putnam–Perigree, New York, 1981.

Dworkin, Andrea, *Right-Wing Women: The Politics of Domesticated Females*, The Women's Press, London, 1983.

Dworkin, Andrea, *Intercourse*, Free Press, Macmillan, New York, 1987.

Easlea, Brian, *Fathering the Unthinkable: Masculinity, Scientists and the Nuclear Arms Race*, Pluto Press, London, 1983.

Easlea, Brian, *Science and Sexual Oppression: Patriarchy's Confrontation with Woman and Nature*, Weidenfeld & Nicolson, London, 1981.

Ehrenreich, Barbara, *The Hearts of Men: American Dreams and the Flight from Commitment*, Pluto Press, London, 1983.

Ehrenreich, Barbara, and English, Dierdre, *For Her Own Good: 150 Years of the Experts' Advice to Women*, Doubleday, New York, 1979.

Eisenstein, Zilla (ed.) *Capitalist Patriarchy and the Case for Socialist Feminism*, Monthly Review Press, New York, 1979.

Eisenstein, Hester, *Contemporary Feminist Thought*, Allen & Unwin, Sydney, 1984.

Evans, Mary (ed.), *The Woman Question: Readings on the Subordination of Women*, Fontana, Oxford, 1982.

Faderman, Lillian, *Surpassing the Love of Men: Romantic Friendship and Love Between Women From the Renaissance to the Present*, Morrow, New York, 1981.

Faulkner, Wendy, and Arnold, Eric (eds), *Smothered by Invention: Technology in Women's Lives*, Pluto Press, London, 1985.

Ferrier, Carole, 'Introductory Commentary: Women Writers in Australia', in Carole Ferrier (ed.), *Gender, Politics and Fiction. Twentieth Century Australian Women's Novels*, University of Queensland Press, Brisbane, 1985.

Firestone, Shulamith, *The Dialectic of Sex: The Case for Feminist Revolution*, Paladin, London, 1972.

Freeman, Jo *Women: A Feminist Perspective*, 2nd edn, Mayfield, Palo Alto, California, 1979.

Friedman, Scarlet, and Sarah, Elizabeth (eds), *On the Problem of Men: Two Feminist Conferences*, Women's Press, London, 1982.

Frieze, Irene, et al., *Women and Sex Roles: A Social Psychological Perspective*, Norton, New York, 1978.

Gale, Fay (ed.), *We Are Bosses Ourselves: The Status and Role of Aboriginal Women Today*, Australian Institute of Aboriginal Studies, Canberra, 1983.

Gage, Matilda Joslyn, *Woman, Church and State* (1893), Persephone Press, Massachusetts, 1980.

Game, Rosemary, and Pringle, Ann, *Gender at Work*, Allen & Unwin, Sydney, 1983.

Gardiner, Judith Kegan, 'On Female Identity and Writing by Women', *Critical Inquiry* 8, 2, 1981, pp. 347−61.

Gilbert, Sandra, '"My Name is Darkness": The Poetry of Self-definition', *Contemporary Literature* XVIII, 4.

Gilbert, Sandra, and Gubar, Susan, *The Madwoman in the Attic: Re-visions of Nineteenth-century Women Writers*, Yale University Press, New Haven, 1979.

Gilman, Charlotte Perkins, *The Man-made World or, our Androcentric Culture*, Charlton, New York, 1911.

Greene, Gayle, and Kahn, Coppélia (eds), *Making a Difference: Feminist Literary Criticism*, Methuen, London and New York, 1985.

Grieve, Norma, and Grimshaw, Patricia (eds), *Australian Women: Feminist Perspectives*, Oxford University Press, Melbourne, 1981.

Griffin, Susan, *Pornography and Science: Culture's Revenge against Nature*, Women's Press, London, 1981.

Grimshaw, Patricia, and Strahan, Lynne (eds), *The Half-Open Door: Sixteen Modern Australian Women Look at Professional Life and Achievement*, Hale & Iremonger, Sydney, 1982.

Harding, Sandra, *The Science Question in Feminism*, Open University Press, Milton Keynes, 1986.

Hartstock, Nancy C. M., *Money, Sex and Power: Toward a Feminist Historical Materialism*, North Eastern University Press, Boston, 1985.

Henley, Nancy, *Body Politics: Power, Sex and Non-Verbal Communication*, Prentice-Hall, New Jersey, 1977.

Holmes, Helen B, Hoskins, Betty B., and Gross, Michael (eds), *The Custom-Made Child? Women-Centred Perspectives*, Humana Press, New Jersey, 1981.

Hooks, Bell, *Feminist Theory: From Margin to Center*, South End Press, Boston, 1984.

Hrdy, Sarah Blaffer, *The Woman That Never Evolved*, Harvard University Press, Cambridge, Massachusetts, 1982.

Hubbard, Ruth, 'Reflections on the Story of the Double Helix', *Women's Studies International Quarterly* 2, 3, 1979, pp. 261–74.

Huizer, Gerrit, and Mannheim, Bruce (eds), *The Politics of Anthropology: From Colonization and Sexism Toward a View from Below*, Mouton, The Hague, 1979.

Iseman, Kay, 'Barbara Baynton: Woman as "The Chosen Vessel"', *Australian Literary Studies*, 20th anniversary issue, 1983, pp. 25–37.

Jacobus, Mary (ed.), *Women Writing and Writing about Women*, Croom Helm, London, 1979.

Jaggar, Alison M., *Feminist Politics and Human Nature*, Harvester Press, Sussex, 1983.

Janeway, Elizabeth, *Powers of the Weak*, Knopf, New York, 1980.

Janssen-Jurreit, Marie Louise, *Sexism: The Male Monopoly on History and Thought*, translated by Verne Moberg, Pluto Press, London, 1982.

Jeffreys, Sheila, *The Spinster and Her Enemies: Feminism and Sexuality 1880–1930*, Pandora Press/Routledge & Kegan Paul, London, 1985.

Jeffreys, Sheila, ' "Free from All Uninvited Touch of a Man": Women's Campaigns Around Sexuality, 1880–1914', *Women's Studies International Forum* 6, 5, 1982, pp. 629–46.

Juhasz, Suzanne, 'The Critic as Feminist: Reflections on Women's Poetry, Feminism, and the Art of Criticism', *Women's Studies* 5, 1977, p. 113–27.

Kaplan, Alexandra, and Sedney, Mary Anne, *Psychology and Sex Roles: An Androgynous Perspective*, Little, Brown, Boston, 1980.

Kimball, Gayle (ed.), *Women's Culture: The Renaissance of the Seventies*, Methuen, New Jersey and London, 1981.

Koedt, Anne, Levine, Ellen, and Rapone, Anita (eds), *Radical Feminism*, Quadrangle/New York Times, New York, 1973.

Laws, Sophie, Hey, Valerie, and Eagan, Andrea, *Seeing Red: The Politics of Pre-Menstrual Tension*, Hutchinson, London, 1985.

Lederer, Laura (ed.), *Take Back the Night: Women on Pornography*, Morrow, New York, 1980.

Leghorn, Lisa, and Parker, Katherine, *Women's Worth: Sexual Economics and the World of Women*, Routledge & Kegan Paul, Boston, 1981.

Linden, Robin R., Pagano, Darlene R., Russel, Diana E. H., Star, Susan Leigh (eds), *Against Sadomasochism: A Radical Feminist Analysis*, Frog In The Well, California, 1982.

Lipman-Blumen, Jean, *Gender Roles and Power*, Prentice-Hall, New Jersey, 1984.

Lips, Hilary, and Colwill, Nina Lee, *The Psychology of Sex Differences*, Prentice-Hall, New Jersey, 1978.

Lloyd, Genevieve, *The Man of Reason: 'Male' and 'Female' in Western Philosophy*, Methuen, London 1984.

Lorber, Judith, *Women Physicians. Careers, Status, and Power*, Tavistock, New York and London, 1984.

MacKinnon, Catharine, 'Feminism, Marxism, Method, and the State: An Agenda for Theory', *Signs: Journal of Women in Culture and Society* 7, 3, 1982, pp. 515−44.

Marciano, Linda ('Lovelace'), *Ordeal: An Autobiography*, Citadel Press, 1980.

Marks, Elaine, and de Courtivron, Isabelle (eds), *New French Feminisms: An Anthology*, Harvester, Brighton, 1980.

Mattews, Jill Julius, *Good and Mad Women: The Historical Construction of Femininity in Twentieth Century Australia*, Allen & Unwin, Sydney, 1984.

McConnell-Ginet, Sally, Borker, Ruth and Furman, Nelly (eds), *Women and Language in Literature and Society*, Praeger, New York, 1980.

McInherny, Frances, 'Miles Franklin, *My Brilliant Career* and the Female Tradition', *Australian Literary Studies* 9, 3, 1980.

McNaron, Toni A. H. (ed.), *The Sister Bond: A Feminist View of a Timeless Connection*, Pergamon, Oxford, 1985.

Mercer, Jan (ed.), *The Other Half: Women and Australian Society*, Penguin, Melbourne, 1975.

Miller, Casey, and Swift, Kate, *Words and Women: New Language in New Times*, Anchor/Doubleday, New York, 1976.

Millett, Kate, *Sexual Politics*, Abacus, London, 1972.

Modjeska, Drusilla, *Exiles at Home: Australian Women Writers, 1925−1945*, Angus & Robertson, Sydney, 1981.

Moi, Toril, *Sexual/Textual Politics: Feminist Literary Theory*, Methuen, London and New York, 1985.

Montefiori, Janet, 'Feminist Identity and the Poetic Tradition', *Feminist Review* 13, February 1983, pp. 69−84.

Morgan, Robin, *Sisterhood is Global*, Anchor/Doubleday, New York, 1984.

Morgan, Robin, *The Anatomy of Freedom: Feminism, Physics, and Global Politics*, Anchor/Doubleday, New York, 1984.

Nestor, Pauline, *Female Friendships and Communities: Charlotte Brontë, George Eliot, Elizabeth Gaskell*, Clarendon Press, Oxford, 1985.

Newman, Louise Michele (ed.), *Men's Ideas/Women's Realities. Popular Science, 1870−1915*, Pergamon Press, New York, 1985.

Oakley, Ann, *Women's Work: The Housewife, Past and Present*, Pantheon, New York, 1974.

Oakley, Ann, *The Captured Womb: A History of the Medical Care of Pregnant Women*, Basil Blackwell, Oxford, 1984.

O'Brien, Mary, *The Politics of Reproduction*, Routledge & Kegan Paul, New York, 1981.

Olsen, Tillie, *Silences*, Virago, London, 1980.

Ostriker, Alicia, 'The Thieves of Language: Women Poets and Revisionist Myth Making', *Signs: Journal of Women in Culture and Society* 8, 1, 1982, pp. 68−90.

Parker, Rozsika and Pollock, Griselda, *Old Mistresses: Women, Art and Ideology*, Routledge & Kegan Paul, London, 1981.

Penman, Robin, and Stolk, Yvonne, *Not the Marrying Kind: Single Women in Australia*, Penguin, Melbourne, 1983.

Petchetsky, Rosalind, 'Workers, Reproductive Hazards and the Politics of Protection: An Introduction', *Feminist Studies* 5, 2, 1979, pp. 233–45.

Pollock, Scarlet, and Sutton, Jo, 'Fathers' Rights, Womens' Losses', *Women's Studies International Forum* 8, 6, 1985, pp. 593–600.

Raymond, Janice, *The Transsexual Empire: The Making of the She-Male*, Beacon Press, Boston, 1979.

Raymond, Janice, *A Passion for Friends: Toward a Philosophy of Female Affection*, Beacon Press, Boston, 1986.

Reiter, Rayna, R. (ed.), *Toward An Anthropology of Women*, Monthly Review Press, New York, 1975.

Rhodes, Dusty, and McNeill, Sandra (eds), *Women Against Violence Against Women*, Only Women Press, London, 1985.

Rich, Adrienne, *Of Woman Born: Motherhood as Experience and Institution*, Virago, London, 1982.

Rich, Adrienne, *On Lies, Secrets, and Silence: Selected Prose 1966–1978*, Virago, London, 1980.

Roberts, Helen (ed.), *Doing Feminist Research*, Routledge & Kegan Paul, London, 1981.

Roe, Jill, 'The Significant Silence: Miles Franklin's Middle Years', *Meanjin* 39, 1, 1980, pp. 48–59.

Rohrbaugh, Joanna Bunker, *Woman: Psychology's Puzzle*, Harvester, Brighton, 1980.

Rosaldo, Michelle Zimbalist and Lamphere, Louise (eds), *Woman, Culture and Society*, Stanford University Press, Stanford, California, 1974.

Rosenberg, R, *Beyond Separate Spheres: Intellectual Roots of Modern Feminism*, Yale University Press, New Haven, 1982.

Rothman, Barbara Katz, *The Tentative Pregnancy: Prenatal Diagnosis and the Future of Motherhood*, Penguin, New York, 1987.

Rowbotham, Sheila, *Woman's Consciousness, Man's World*, Penguin, Harmondsworth, 1973.

Rowland, Robyn, 'Technology and Motherhood: Reproductive Choice Reconsidered', *Signs: Journal of Women in Culture and Society* 12, 3, 1987, pp. 512–28.

Rowland, Robyn, 'Women as living laboratories: the new reproductive technologies', in Josefina Figueria-McDonough and Rosemary Sarri (eds), *The Trapped Woman: Catch 22 in Deviance and Control*, Sage, Newbury Park, 1987.

Rowland, Robyn. (ed.) *Women Who Do and Women Who Don't, Join The Women's Movement*, Routledge & Kegan Paul, London, 1984.

Rowland, Robyn, 'Women's Studies Courses: Pragmatic and Political Issues Concerning their Establishment and Design', *Woman's Studies International Forum* 5, 5, 1982, pp. 487–95.

Ruth, Sheila, *Issues in Feminism: A First Course in Women's Studies*, Houghton Mifflin, Boston, 1980.

Ryan, Edna and Conlan, Anne, *Gentle Invaders*, Nelson, Melbourne, 1975.

Sayre, Anne, *Rosalind Franklin and DNA*, Norton, New York, 1975.

Scully, Diana, *Men Who Control Women's Health: The Miseducation of Obstetricians/Gynaecologists*, Houghton Mifflin, Boston, 1980.

Seaman, Barbara, and Seaman, Gideon, *Women and the Crisis in Sex Hormones*, Rawson, New York, 1977.

Showalter, Elaine, *A Literature of Their Own: British Women Novelists from Brontë to Lessing*, Princeton University Press, Princeton, New Jersey, 1977.

Smith-Rosenberg, Carroll and Rosenberg, Charles, 'The Female Animal: Medical and Biological Views of Woman and her Role in Nineteenth-century America', *Journal of American History* LX, 2, 1973.

Smith-Rosenberg, Carroll, 'The Female World of Love and Ritual: Relations between Women in Nineteenth Century America', *Signs: Journal of Women in Culture and Society* 1, 1, 1975.

Snyder, Eloise C. (ed.), *The Study of Women: Enlarging Perspectives of Social Reality*, Harper & Row, New York, 1979.

Spacks, Patricia Meyer, *The Female Imagination*, Avon Books, New York, 1975.

Spender, Dale, *Mothers of the Novel*, Pandora Press/Routledge & Kegan Paul, London and New York, 1986.

Spender, Dale, *Women of Ideas and What Men Have Done to Them*, Routledge & Kegan Paul, London and Boston, 1982.

Spender, Dale, *Man Made Language*, Routledge & Kegan Paul, London, 1980.

Spender, Dale, and Spender, Lynne, *Scribbling Sisters*, Hale & Iremonger, Sydney, 1984.

Spender, Dale (ed.), *Men's Studies Modified: The Impact of Feminism on the Academic Disciplines*, Pergamon, New York and London, 1981.

Spender, Lynne, *Intruders on the Rights of Men: Women's Unpublished Heritage*, Pandora Press/Routledge & Kegan Paul, London and Boston, 1983.

Summers, Anne, *Damned Whores and God's Police: The Colonization of Women in Australia*, Penguin, Melbourne, 1975.

Sydney Women Against Incest, *Breaking The Silence*, a report based upon the findings of the Women Against Incest phone-in survey, Sydney, 1985.

Tavris, Carol, and Offir, Carole, *The Longest War. Sex Differences in Perspective*, Harcourt Brace Jovanovich, New York, 1977.

Tilly, Louise and Scott, Joan, *Women, Work and Family*, Holt, Rinehard and Winston, New York, 1978.

Trask, Haunani-Kay, *Eros and Power: The Promise of Feminist Theory*, University of Pennsylvania Press, Philadelphia, 1986.

Trebilcot, Joyce (ed.), *Mothering: Essays in Feminist Theory*, Rowman & Allanheld, Totowa, New Jersey, 1984.

Unger, Rhoda, *Female and Male: Psychological Perspectives*, Harper & Row, New York, 1979.

Walker, Shirley (ed.), *Who is She? Images of Woman in Australian Fiction*, University of Queensland Press, Brisbane, 1983.

Walum, Laurel Richardson, *The Dynamics of Sex and Gender: A Sociological Perspective*, Rand McNally, Chicago, 1977.

Ward, Elizabeth, *Father—Daughter Rape*, Women's Press, London, 1984.

Weideger, Paula, *Menstruation and Menopause: The Physiology and Psychology, the Myth and the Reality*, Penguin, New York, 1977.

Weitz, Shirley, *Sex Roles: Biological, Psychological and Social Foundations*, Oxford University Press, New York, 1977.

Windschuttle, Elizabeth (ed.), *Women, Class and History: Feminist Perspectives on Australia 1788—1978*, Fontana, Sydney, 1980.

Index